INTERNATIONAL UNION
OF GEOLOGICAL SCIENCES

PUBLICATION NO.13

SOUTHEAST ASIA: TECTONIC FRAMEWORK, EARTH RESOURCES AND REGIONAL GEOLOGICAL PROGRAMS

by

John A. Katili and John A. Reinemund

July 1, 1984

INTERNATIONAL UNION OF GEOLOGICAL SCIENCES

E. Seibold (FRG), President
C. C. Weber (France), Secretary General
J. A. Reinemund (USA), Treasurer

D. F. Merriam (USA), Chairman, Advisory Board on Publications
L. Hoover (USA), Editor-in-Chief

— — —

The International Union of Geological Sciences (IUGS) was established in 1961 to promote and encourage the study of geological problems, especially those of world-wide significance. With a current membership of 91 countries, it is now one of the world's largest and most active nongovernmental scientific organizations. IUGS supports and facilitates international cooperation in the geological sciences and is the scientific sponsor of the quadrennial International Geological Congress. With UNESCO, it sponsors the International Geological Correlation Program, and it is a co-sponsor, with the International Union of Geodesy and Geophysics, of the International Lithosphere Program.

— — —

Copyright © 1984, IUGS
ISBN 0-930423-08-9

This publication is available from IUGS, Room 177, 601 Booth Street, Ottawa, Ontario, Canada K1A OE8, and from IUGS Secretariat, Department of Geology, Norwegian Institute of Technology, 7034-NTH Trondheim, Norway.

INTERNATIONAL UNION OF GEOLOGICAL SCIENCES

PUBLICATION NO. 13

SOUTHEAST ASIA: TECTONIC FRAMEWORK, EARTH RESOURCES
AND REGIONAL GEOLOGICAL PROGRAMS

Highlights and results of a seminar on Development and Coordination
of Regional Geological Programs in Developing Countries held in
Bangkok, Thailand, January 24-25, 1983

by

John A. Katili and John A. Reinemund

July 1984

Acronyms

AGID	Association of Geoscientists for International Development
ASCOPE	ASEAN Council on Petroleum
ASGA	Association des Services Geologiques Africains/ Association of African Geological Surveys
CCOP/EA	ESCAP Committee for Co-ordination of Joint Prospecting for Mineral Resources in East Asian Offshore Areas.
CCOP/IO	ESCAP Committee for Co-ordination of Joint Prospecting for Mineral Resources in Indian Ocean Offshore Areas.
CCOP/SOPAC	ESCAP Committee for Co-ordination of Joint Prospecting for Mineral Resources in South Pacific Offshore Areas
CERESIS	Centro Regional de Seismologia para America del Sur/Regional Center for Seismology of South America
CGMW	IUGS Commission for the Geological Map of the World
COGEODATA	IUGS Committee on Storage, Automatic Processing, and Retrieval of Geological Data
CPCEMR	Circum-Pacific Council for Energy and Mineral Resources
CPMP	Circum-Pacific Map Project
ESCAP	UN Economic and Social Commission for Asia and the Pacific
GEOSEA	Regional Conference on the Geology and Mineral Resources of Southeast Asia
GSA	Geological Society of Africa
IDOE	International Decade of Ocean Exploration
IGCP	IUGS/UNESCO International Geological Correlation Programme
IOC	Intergovernmental Oceanographic Commission
IUGS	International Union of Geological Sciences
PAIGH	Pan American Institute for Geography and History
RMRDC	ESCAP Regional Mineral Resources Development Centre
SEAPEX	South East Asia Petroleum Exploration Society
SEATAR	Studies of East Asia Tectonics and Resources (CCOP/IOC)
SEATRAD	Southeast Asian Tin Research and Development Centre
UNESCO	United Nations Educational, Scientific and Cultural Organization.

CONTENTS

PART I: DEVELOPMENT AND COORDINATION OF REGIONAL GEOLOGICAL PROGRAMS IN SOUTHEAST ASIA, by John A. Reinemund

	Page
Purpose and organization	1
Background and objectives	1
Scope and agenda	2
Participating organizations	2
Facilities and arrangements	3
Presentations concerning international programs	3
ESCAP programs in Southeast Asia	3
Other programs in Southeast Asia	7
Regional programs in Africa and Latin America	9
Discussions of issues, problems, and mechanisms	9
New initiatives and program direction	9
Issues and problems	12
Summary and conclusions	14
List of seminar participants	15

PART II: TECTONIC FRAMEWORK, RESOURCES OCCURRENCES AND RELATED PROBLEMS IN SOUTHEAST ASIA, by John A. Katili

Introduction	18
Broad tectonic framework of Southeast Asia	19
Cenozoic tectonic framework	19
Mesozoic tectonic framework	25
Paleozoic tectonic framework	28
Precambrian tectonic framework	32
Tectonic evolution of Southeast Asia	32

Fundamental problems of and contradicting views on the plate tectonics of Southeast Asia ... 39

    Types of subduction ... 39

    Strongly curved arcs ... 41

    Seismicity and form of the Banda Arc ... 41

    The age of the Banda Basin ... 42

    Rotation of Sulawesi ... 42

    Opening of the Makassar Strait ... 43

    Position of Southeast Asia in relation to continental drift ... 43

    Rotation of Sumatra ... 43

Tectonic environment of the Southeast Asian mineral and hydrocarbon deposits ... 44

    Mineral deposits and tectonic framework ... 44

    Basin setting and tectonic framework ... 54

General conclusions and recommendations for future research ... 59

Acknowledgements ... 63

REFERENCES ... 64

Part I: DEVELOPMENT AND COORDINATION OF REGIONAL GEOLOGICAL
PROGRAMS IN SOUTHEAST ASIA

by
John A. Reinemund

PURPOSE AND ORGANIZATION

Background and objectives

In order to encourage the participation of developing countries in international geoscience programs, the Executive Committee of the International Union of Geological Sciences (IUGS) holds its annual meeting in alternate years in one of the developing countries belonging to the Union. It also sponsors a seminar each year, immediately preceding the Executive Committee meeting, on a subject of importance in planning future international programs.

The 1983 IUGS Executive Committee meeting was held at the Asian Institute of Technology near Bangkok, Thailand, January 26-29. It was preceded on January 24-25 by a seminar on "Development and Coordination of Regional Geological Programs in Developing Countries," held at the headquarters of the U.N. Economic and Social Commission for Asia and the Pacific (ESCAP). The seminar was jointly sponsored by IUGS, ESCAP, the Association of Geoscientists for International Development (AGID), the ESCAP Coordinating Committee for Joint Prospecting of Mineral Resources in East Asian Offshore Areas (CCOP/EA)*, and UNESCO. This report summarizes the discussions and reviews the conclusions of the seminar.

IUGS has long been concerned about increasing the participation of developing country organizations and scientists in international geoscience programs. Such participation is important for the worldwide study of geological problems of concern to IUGS. It also contributes to strengthening the geological organizations and programs in these countries, through interchange with the international geoscience community.

It has been apparent, however, that strong programs of regional cooperation in geology can be helpful in facilitating the participation of developing countries in geological programs of worldwide scope, as well as in coordinating national geological programs. Such regional programs have been especially effective in Southeast Asia. Because of this, IUGS decided to sponsor a seminar to examine the mechanisms for regional geological cooperation, with special reference to Southeast Asia, attempting to draw conclusions that might be more widely applicable. In his opening remarks, at the seminar, Prof. E. Seibold, President of IUGS, expressed the Union's interest as follows:

> "IUGS must promote advancement of science, and in this region we have tried to do so together with many other organizations. So far, many

---

* Also known simply as CCOP

programs and projects have been very successful. Why? Perhaps one of the reasons is the long cultural tradition of this region. Perhaps it is the very urgent need to understand better your friendly but often threatening natural environment. Perhaps it is your wealth in mineral resources to be explored and exploited, and our general concern to restrict environmental damage.

"During the next few days we are asking you why these programs have been so successful, what was achieved, and what is left to do. What factors of international cooperation have been especially helpful? Have we been able to raise the interest of the very best scientists in this historical task, to help the development of this fascinating region located between a vast and young ocean and an old continent?"

Scope and agenda

In an attempt to find answers to the questions posed by Prof. Seibold, the seminar discussions were directed toward the following specific objectives:

1. To focus attention on the mechanisms for international geoscience cooperation, especially in Southeast Asia.

2. To examine the effectiveness of these mechanisms in relation to resource problems and needs for cooperation.

3. To identify possibilities for wider use of these mechanisms and increased cooperation between IUGS and geoscience organizations in Southeast Asia.

As background for the discussions and in order to obtain a perspective on the resources and problems of Southeast Asia, IUGS asked Dr. J. A. Katili of Indonesia to present an overview of the geologic framework of the region as related to known and potential resources. His overview is published as Part 2 of this report. Thereafter, brief presentations of functions, activities, and problems of regional geoscience programs were given by representatives of most intergovernmental and international scientific organizations involved in Southeast Asia. This was followed on the second day of the seminar by brief presentations on programs in Africa and Latin America, and then by discussions of new and innovative programs. Finally, the presentations were reviewed by a panel, which made comments and recommendations on principal issues that were considered during the seminar.

Participating organizations and persons

A wide range of governmental and nongovernmental organizations participated in the seminar. As reported in Episodes (1983/1, p. 38), these organizations "range widely from governmental agencies like ESCAP and UNESCO to nongovernmental organizations such as the Circum-Pacific Map Project (CPMP) and the Commission for the Geological Map of the World (CGMW). Some function largely as administrative and coordinating bodies (e.g. ESCAP Mineral Resources Section), while others like RMRDC (ESCAP's Regional Minerals Resources Development Centre and CCOP are actively carrying out field projects. There are those such as CCOP,

GEOSEA (the triennial Regional Conference on the Geology and Mineral Resources of SE Asia), and IGCP that are well established and widely recognized as successful and those that are younger and less well known such as the Southeast Asian Tin Research and Development Centre (SEATRADC)." The names and organizational affiliations of those persons who attended the seminar are shown on pages 16 and 17.

Facilities and arrangements

Facilities for the seminar were provided by ESCAP, through the generosity of S.A.M. Kibria, Executive Secretary of ESCAP, who acted as host for the seminar. Logistical arrangements were coordinated by an interagency steering committee, consisting of Prinya Nutalaya of the Asian Institute of Technology, L. Machesky of ESCAP, and E. Dubois of CCOP/EA. IUGS and the cosponsoring organizations appreciate the efforts of these individuals and their staffs. The assistance of A.R. Berger, M.G. Bassett, Z. Kielan-Jaworowska, and J.V. Hepworth, who served as rapporteurs for the discussions, is also greatly appreciated.

PRESENTATIONS CONCERNING INTERNATIONAL PROGRAMS

ESCAP programs in Southeast Asia

Many of the regional programs in Southeast Asia are conducted within the framework of the Natural Resources Division of ESCAP. The activities of that Division, which are directed toward the evaluation, development, and use of mineral, water, and energy resources, were presented by L. Machesky as functions of 10 organizations or programs, as decribed below.

1) Mineral Resources Section

The priority program of the Mineral Resources Section of ESCAP, based in Bangkok, involves collation and analysis of relevant data from member countries in order to assist the ESCAP countries in strengthening their capabilities for planning exploration and development of mineral and energy resources. The program elements are 1) review and analysis of mineral exploration and development in the region, 2) regional geological and thematic mapping, 3) review and compilation of geology of the region, and 4) studies of the mineral potential of the region. Other activities include technical and administrative assistance to regional projects: the Regional Mineral Resources Development Center (RMRDC) at Bandung, Indonesia; the Southeast Asia Tin Research and Development Centre (SEATRADC) at Ipoh, Malaysia; CCOP/South Pacific (SOPAC) at Suva, Fiji; CCOP/East Asia (EA) at Bangkok; and CCOP/Indian Ocean (IO), currently being established.

The exploration and development program is centered around resources that have considerable impact on the economies of the countries in the region, such as oil, natural gas, coal, tin, copper, nickel, fertilizers, and construction materials.

There is a recognized need for all countries to have an adequate and usable technical data base that would provide to planners and specialists evidence of the presence or absence of mineral resources, and to scientists the background information to enable them to better understand the geology and the processes leading to the formation and emplacement of economic mineral resources. ESCAP contributes to this data base by publication of scientific maps and technical publications like the "Mineral Resources Development Series", of which 48 issues have been published to date.

A number of regional geological and other specialized maps have been compiled and published under the sponsorship of ESCAP since 1959. The latest is the second edition of the Mineral Distribution Map of Asia, published in 1980. The third edition of the Oil and Gas Map of Asia is under preparation.

The project on stratigraphic correlation among sedimentary basins of the ESCAP region was initiated in 1958 and was subsequently merged with IGCP Project 32. The major emphasis of the project is to depict the nature, structure, age, thickness, facies, and sedimentary sequences in the depositional basins of the ESCAP region. The purpose is to further enrich the knowledge of tectonic and paleogeographic controls of mineral and chemical distribution in the Earth's crust. The project includes the compilation of an atlas of sedimentary basins of the ESCAP region and the correlation between them. The basins are also analyzed in relation to hydrocarbon occurrences, their origin and distribution, with a view to recognizing the factors in sedimentary basin development that favored the generation and accumulation of mineral and energy resources. Such factors have not only theoretical but also practical significance for countries of the region in their efforts to develop their natural resources. Although Project 32 was to formally end by the end of 1982, the member countries' representatives agreed in November 1982 that the work should be continued, even if the IGCP does not extend the Project.

ESCAP also promotes technologies for exploration and development of mineral resources in the developing countries of the region. It organizes seminars and study tours which are among the most effective forms of training and exchange of scientific and technological information. Examples are the seminar and study tour on modern methods of mineral prospecting, which was held in 1980 in the USSR, and a seminar on drilling, sampling, and borehole logging, which was also held in the USSR in 1981.

2) RMRDC

The Regional Mineral Resources Development Centre, based in Bandung, Indonesia, (RMRDC) has a staff of experts assigned by ESCAP cooperating countries to assist ESCAP member countries with specific investigations and problems (Hepworth, 1982). These experts have carried out over 200 technical and advisory missions to 23 member and associate member countries. Some of the tasks of the Centre are to establish a network of specialized laboratories such as age-dating centres, to organize large-scale airborne geophysical surveys, and to organize training for the scientists and technologists in the member countries.

3) SEATRADC

The South East Asia Tin Research and Development Centre (SEATRADC), based in Ipoh, Malaysia, is an intergovernmental organization established by Indonesia, Malaysia and Thailand in 1978, with the assistance of ESCAP and UNDP.

The Centre's research and development programme envisages exploration of tin in member countries, development of drilling and sampling techniques, development of geostatistical methods in tin ore-reserve calculations, improvement of hydraulic mining techniques, mined-land rehabilitation, improvement of recovery of tin from smelters in the region, and development of a documentation unit for the collection and dissemination of information. Fellowships for training abroad, and in-service training at the Centre and other related research institutes in the region, are made available to national staff. The Centre also organizes group training in the form of workshops, study tours, and seminars.

4) CCOP/EA

The highly successful Committee for Co-ordination of Joint Prospecting for Mineral Resources in East Asian Offshore Areas (CCOP/EA) was established by the member Governments, under the aegis of ESCAP, in 1966. It has been providing consultancy and field services, coordinating joint surveys and research projects, and looking after the training requirements of the member countries. During its 16 years of service it has made tangible contributions in the field of exploration and development of offshore petroleum and mineral resources and has generated and disseminated substantial amounts of basic geoscientific data.

The Committee issues a quarterly newsletter on geoscience and resources developments in the region, and, with the support of Japan, publishes reports on the results of investigations in the region. CCOP/EA has sponsored workshops on subjects related to resources exploration and assessment, and has carried out a program for introducing petroleum data systems in cooperation with the IUGS Committee on Storage, Automatic Processing, and Retrieval of Geologic Data (COGEODATA). CCOP/EA has also had an effective program of assistance to member countries in offshore geophysical surveys for detrital minerals.

5) Regional Centre for Quaternary Geology

In recognition of the identified needs of many of its member countries to undertake investigations and research by adequately trained professional/technical staff, CCOP/EA, with assistance from ESCAP, has worked toward establishing a regional center for training and research in Quaternary geology. At its nineteenth session at Tokyo in late 1982, CCOP/EA accepted the offer of China to provide the host facilities at Qingdao under the basic premise that the Centre would be open to CCOP countries and to all member and associate member countries of ESCAP. It is expected that the Centre will provide training in Quaternary geology and related subjects, provide experts for training and advisory services to participating countries, and act as a documentation and information centre.

6) CCOP/SOPAC

The intergovernmental CCOP for South Pacific Offshore Areas (CCOP/SOPAC), based in Suva, Fiji, has been doing valuable work since its inception in 1972.

CCOP/SOPAC, whose member countries are Cook Islands, Fiji, Kiribati, New Zealand, Papua New Guinea, Samoa, Solomon Islands, Kingdom of Tonga, and Vanuatu, has the following objectives: to discover and develop offshore mineral and hydrocarbon deposits; to reconnoiter and identify economically exploitable deposits; to either execute or promote execution of more detailed surveys and studies where desirable; to acquire basic physical and chemical data required for engineering and other development work in the coastal zones; to execute specific offshore surveys, together with marine baseline data for evaluation of energy potential and for formulation of design criteria for engineering and other developmental works in the coastal zone; and to train member country nationals.

7) CCOP/IO

Establishment of a Committee for Indian Ocean Exploration, based on the expressed needs and capabilities of the countries concerned, has been under consideration by the Governments of Bangladesh, India, and Sri Lanka. Pakistan and Maldives have also expressed their interest in the Committee. Other possible members are Burma and Iran. The envisaged functions of the Committee are similar to those of CCOP/EA and CCOP/SOPAC.

8) Water Resources Section

The priority programs of the ESCAP Water Resources Section are designed to assist member countries in the use of their available water resources; to draw up accurate water-demand projections using low-cost equipment, processes and facilities; to improve systems for data on water use; to formulate and establish appropriate national policies as well as institutional and legal arrangements; to develop a core of skilled staff with the capacity to apply water resources planning techniques; to promote intergovernmental cooperation; to determine the most efficient combination of energy production requirements with minimum water use; to assess damage from cyclones, floods, and droughts, and plan and carry out measures to mitigate this damage; to maintain an awareness of developments of interest in water resources matters; and to provide advisory services on request.

9) Energy Resources Section

The main tasks of the Energy Resources Section are to collect and dissemnate information on the availability, use, supply and demand of energy resources; to assist member countries in formulating and carrying out integrated programs for the development, use, and management of their energy resources, including consideration of legal and institutional matters and financial requirements; and to foster the production and use of appropriate forms of energy and appropriate energy mix in rural areas, utilizing, as much as possible, energy resources, materials, and suitable labor available.

10) Remote Sensing, Surveying and Mapping

This program is planned to promote the cooperative search for solutions to common problems in the ESCAP countries in the field of natural resources development and management, including monitoring of the environment, by strengthening country capabilities in the use of remote sensing technology, methods, and techniques. A regional remote sensing project with funding of about $US 1.5 million from UNDP and with support from participating and donor countries is being implemented in 1983.

In the discussions of these ESCAP programs it was pointed out that one of the principal strengths of both CCOP/EA and CCOP/SOPAC are their Technical Advisory Groups, which meet annually along with each Committee, review the technical program, recommend future activity, and help each Committee achieve its objectives. This is a unique and powerful mechanism for encouraging and facilitating these activities and for making certain that appropriate scientific concepts and advanced technology are applied in the respective programs.

Other programs in Southeast Asia

In addition to the ESCAP programs, Southeast Asia has substantial activity under other intergovernmental and nongovernmental programs. Some of these were reviewed during the first day of the seminar, as summarized below.

1) UNESCO Regional Office for Science and Technology

The UNESCO regional office in Jakarta facilitates and coordinates scientific and technological activities supported by UNESCO in Southeast Asia. These include regional seismological studies, backup for several continuing UNESCO programs, and support for selected workshop, training, or research activities in the earth sciences.

2) International Geological Correlation Program (IGCP)

IGCP is a joint UNESCO/IUGS program, and several of its projects involve activities in Southeast Asia. Project 32, on the stratigraphy of sedimentary basins in the ESCAP region, which has already been mentioned, is a highly successful project initiated by ESCAP. Other projects involving this region include circum-Pacific plutonism, circum-Pacific siliceous sediments, sea-level changes, and acid magmatism. A review of existing and potential IGCP projects in Southeast Asia was undertaken immediately following the seminar.

3) Commission for the Geological Map of the World (CGMW)

CGMW compiles and publishes continental maps of Southeast Asia and adjacent areas through its Subcommission for Asia and the Far East. These activities are coordinated by the Geological Survey of India, and involve inputs from all the Southeast Asian countries. The following maps sponsored by CGMW are either published or in preparation:

Geographic Base Map of South and East Asia, 1:10,000,000 scale, 1 sheet, and 1:5,000,000, 4 sh., 1979, CGMW/Geol. Surv. India.

Metamorphic Map of South and East Asia, 1:10,000,000 scale, in preparation, CGMW.

Geologic Map of Asia and the Far East, 1:500,000 scale, 1971, 2nd edition, 4 sh., Expl. Note., UN-ECAFE/UNESCO.

Geologic Map of South and East Asia, 1:500,000 scale, 3rd edition, in preparation, CGMW/Geol. Surv. India.

Tectonic Map of South and East Asia, 1:500,000 scale, 7 sh., CGMW/Geol. Surv. India.

Metallogenic Map of South and East Asia, 1:500,000 scale, in preparation, CGMW.

4) Circum-Pacific Map Project (CPMP)

The Circum-Pacific Map Project is an activity of the Circum-Pacific Council for Energy and Mineral Resources (CPCEMR) and involves the participation of more than 30 countries in compiling more than 40 geologic, tectonic, and resources maps of the Pacific region. CCOP/EA and CCOP/SOPAC are linked with this Project, which is organized in five panels covering the four quadrants of the Pacific Basin and the Antarctic region. The Northwest Panel is a working group of CCOP/EA and the Southwest Panel is a working group of CCOP/SOPAC. Compilation of the maps is carried out by these panels, coordinated with the needs and interests of the two committees. A new joint project of CPMP, CCOP/EA, and IUGS is being formulated to compile 1:2 million-scale maps and data for sedimentary basins in Southeast Asia in coordination with IGCP Project 32.

5) Southeast Asia Tectonics and Resources program (SEATAR)

The SEATAR program originated as a joint effort of CCOP/EA and the Intergovernmental Oceanographic Commission (IOC), with considerable financial support from the U.S. National Science Foundation. Coordinated offshore and on-land surveys were carried out first along five selected transects, resulting in an outstanding series of bathymetric and geophysical maps, published by the Geological Society of America, plus a series of volumes now being published (e.g. Hayes, 1983). The program is a continuing effort of CCOP/EA, involving a gradually enlarging series of coordinated land-seabed surveys.

6) Geological Conference of Southeast Asia (GEOSEA)

As a regional meeting held every three years and rotated among the Southeast Asia countries, GEOSEA provides an opportunity to review progress of geoscience activities in the region (Tan and Khoo, 1982). Most organizations active in the region use these meetings to hold program planning sessions.

7) Association of Geoscientists for International Development (AGID)

AGID has its headquarters in Southeast Asia and is active in the region. Most recently it held a "Landplan I" conference in Bangkok, dealing with problems of environmental and land-utilization geology. AGID is planning a program on urban geology in cooperation with IUGS and ESCAP.

Regional programs in Africa and Latin America

The second day of the seminar began with brief reviews of regional programs in Africa and Latin America. On both continents, IGCP projects are increasingly important in regional geological research. In Africa, UNESCO is supplementing IGCP with a regional study of Precambrian geology and resources. CGMW has also been active in both continents, having produced geologic and tectonic maps; metallogenic maps are in preparation.

In Africa, the recently-formed Geological Society of Africa (GSA) provides an innovative framework for continental cooperation, supplementing the Association of African Geological Surveys (ASGA), which has long served as a coordinating mechanism. The Economic Commission for Africa has a limited program of geoscience and resource activity, which includes a regional mineral development center in Tanzania. Regional remote sensing training centers have been established in Kenya and Upper Volta.

In South America, the Council of Directors of Geological Surveys is a developing mechanism for continental cooperation. A regional seismological research network and program (CERESIS) has been organized for the Andean countries. Also, the Commission on Geophysics of the Pan American Institute of Geography and History (PAIGH) has for many years provided an effective mechanism for cooperation primarily in seismology and crustal studies. The Southeast Panel of the Circum-Pacific Map Project is an evolving mechanism of cooperation. Moreover, IUGS is becoming more active in Latin America, having recently initiated metallogenic research projects on tin in the Andes region and on mafic/ultramafic rocks in the Caribbean region.

DISCUSSION OF ISSUES, PROBLEMS, AND MECHANISMS

New initiatives and program directions

Following the presentations on Africa and Latin America, discussions of new program possibilities or initiatives revolved around talks by E. DuBois on possible mechanisms in Southeast Asia, by W. W. Hutchison on the IUGS Research Development Program, by F. Hehuwat on scientific research activities, and by W. H. Matthews on geological education and information flow. These talks and subsequent discussions were recorded and summarized by Z. Kielan-Jaworowska as follows:

1) E. DuBois reviewed the Southeast Asian institutions and organizations which are involved in coordination of geosciences in the region: ESCAP, CCOP, IGCP Project 32, RMRDC, SEATRADC, AGID, SEATAR, ASCOPE, and SEAPEX, along with some local variants of CPCEMR, CGMW, IGCP, and the efforts of UNESCO in earth and marine sciences. In spite of the existence of these institutions and organizations, there is a need for an institution of umbrella type which would handle regional and sub-regional investigations in the fields of natural resources and earth science investigations. To this end, Dr. DuBois made a proposal to create an Asian Institute of Earth Sciences and Resources in Bangkok or elsewhere. The institute would be headed by a director and advised by a board of councillors. The work program would be organized in sections on a) resources (hydrology, hydrocarbons, coal and lignite, economic minerals,

construction materials), b) applied science (cartography and remote sensing, engineering geology and seismology, stratigraphy, marine geology, crustal structure, geochronology, geochemistry, and Quaternary geology), and c) services (technical assistance, training, fellowships, publications, library, and computer center.)

Funding for this institute would necessarily come from the regional countries and organizations, and from contributions from cooperating countries and institutions. Initial staffing would be some 20 professionals plus 20 secondments. The budget would be about US$ 2.5 million per year.

Specific objectives of the institute would be to supplement local country programs, to carry out regional programs, and to provide training, advice, and technical assistance. Additional objectives would be to provide an institution to carry on geoscience programs of an international character, provide a continuing pool of experts, and provide training for geoscientists from member countries. It was also suggested that the new institute should work in close cooperation with ESCAP.

Discussion--E. Von Braun, J. V. Hepworth, and J. A. Katili took part in the discussion of Dr. DuBois' proposal. Dr. Von Braun stated that an institute (Regional Mineral Resources Development Center in Bandung) already has been created to this end. Dr. Hepworth said that it is a great idea, but asked why one of the structures which already exists in Southeast Asia cannot play the same role. Similarly, Dr. Katili was of the opinion that it would be impractical to create a super-structure as suggested by Dr. DuBois, as it would duplicate existing organizations. Dr. Machesky agreed with this view.

2) W.W. Hutchison discussed the IUGS Research Development Program, which was established in 1980. The first meeting of the Program Board took place in Bonn in 1982. The functions of the Board are to review ongoing scientific activities of IUGS in order to recommend 1) a possible redirection of effort, 2) the initiation of new programs or activities, and 3) the formulation of possible areas of future research.

One project undertaken by the Board in 1982 illustrates what can be accomplished. In August 1982, a seminar, "Petroleum Resources Assessment," was organized in Honolulu, Hawaii, sponsored by CCOP, CPEMR, the Geological Survey of Canada, IRDC, and the U.S. Geological Survey. The meeting attracted 111 participants from 41 countries. It has been recommended that IUGS establish a standing committee to study this topic.

A proposed project concerns "Sedimentary Basin Analysis." Various techniques used in resource studies may depend on a knowledge of the evolution of sedimentary basins. As geology is not limited by political boundaries, the study of this problem may be done only through international cooperation. In terms of basin analysis, Southeast Asia may play an important role because 1) there has already been extensive international cooperation through CCOP, ASCOPE, ESCAP, and other organizations; 2) much work has been done by CGMW, the Circum-Pacific Map Project, and IGCP Project 32; and 3) outputs are important both economically (water, oil, minerals) and socially (pollution, waste disposal).

IUGS may initiate the project but regional cooperation is vital to its success. IUGS urges Southeast Asian geoscientists to consider strengthening basin analysis studies, which would serve not only as an integral part of a global program, but also as an example of regional international cooperation. The example of this cooperation might be applied elsewhere in other regions of the world.

Discussion--In the discussion, Dr. Katili gave examples of fruitful cooperation of various geoscience organizations in Southeast Asian countries.

3) F. Hehuwat discussed the conditions under which international cooperative research programs can be conducted successfully and pointed out that some of the constraints include the differences between the developed and underdeveloped countries in their scientific objectives and approaches, and the comparative extent of personnel development and economic conditions that affect available resources. He discussed these problems using the example of SEATAR, which is very successful and in which both developed and developing countries participate. SEATAR has been successful because 1) a well defined geological problem provided a common ground for work by scientists from developed and developing countries; 2) the research objectives of developed countries were integrated with mineral resource investigation needs of developing countries; 3) the support of local national governments and mineral industries was obtained by accommodating their needs as well as those of the scientific community; 4) the available resources were concentrated within a reasonably small geographic area; and 5) transfer of technology and expertise was assured by the cooperative research effort and attendant training effects on local investigators. Further cooperation should ensure that the personnel from developing countries should participate as full partners with their associates from the developed countries. Medium and longterm projects with well defined objectives at timely intervals would enhance the benefits to be expected from such cooperative efforts.

Discussion--In the discussion, E. Seibold drew attention to the fact that most of the discussion in this seminar was focused on active parts of the science (calculation of amount of mineral resources, etc.). However, one should not forget the passive part of geology--what geologists can do for society, to protect society, to protect the landscape, to prepare waste disposal programs. In this respect, the geosciences should act also as a sort of insurance company for society.

4) W. H. Matthews, speaking about geoscience education and information flow, gave examples of how education and flow of information are being accomplished in the United States, suggesting that some of the methods used may be adopted in the Southeast Asian region. He pointed out that it is important at the beginning to make the layman familiar with geoscience problems, as well as to introduce the elements of geoscience into pre-college education. For the trained geologist, it is important to have access to geological literature of various spheres, to attend workshops and meetings as well as to have an access to newsletters that should be published in a larger scale than hitherto.

Discussion--J. A. Katili, F. Delaney, J. V. Hepworth, E. Seibold, and E. Von Braun took part in the discussion; the three first speakers gave examples of workshops and field courses organized to train geologists in various countries. Dr. Seibold asked whether there is a special branch in UNESCO for training geologists; Dr. Von Braun answered that some UNESCO fellowships are used for geoscientists.

Issues and problems

In the closing session of the seminar, the deliberations of the previous sessions were summarized by two rapporteurs, A. R. Berger and M. G. Bassett, and reviewed by a panel of seven of the participants. J. V. Hepworth prepared the following summary of the panel discussion.

It was generally agreed that an understanding of the various organizations concerned with the earth sciences in the region had been gained; these include governmental and intergovernmental bodies, international organizations, and regional organizations. A few other organizations in the region were not represented at the meeting. The participants were impressed by the number and diversity of these bodies and the generally effective way in which they functioned.

It was further agreed that Dr. Katili had provided a brilliantly illuminating account of the geology and tectonics of the Southeast Asian region, which gave a sense of geological purpose to the meeting. He had also specified a number of practical problems that need to be solved.

The participants were satisfied that the most important parts of the regional geological program had been described, that the main problems had been recognized, and that solutions to some of the problems could now be considered. The IUGS Executive Committee and all other participants in the seminar had received an insight into what could be recommended as the role of IUGS in encouraging and assisting further development in the region.

Broadly speaking, the issues can be considered under three headings: Science and Research, Program Management, and Education and Communication.

1) It was agreed that the SEATAR program, which had extended from 1973 to the present, based on plans developed in a workshop in Bangkok in 1973, under the auspices of IDOE and IOC, with CCOP as main implementing agent, had been profoundly successful by employing the transect concept to bring about a major understanding of the geology and tectonics of the region. The publication of the "Blue Book," based on the two consultants' 6-month study and with the advice of the Technical Advisory Group and the Bandung meeting's review, was a significant achievement. This might be a model for a similar periodic review, and it was agreed that a "second SEATAR" program should be seriously considered as a major recommendation. However, in such a "SEATAR II," more emphasis should be given to the resource aspect and a greater attempt made to involve petroleum companies in the new program.

If a "SEATAR II" program were developed, every effort should be made to involve national surveys and universities in the transects. Their role might be in "building a bridge" between the marine sections of the transects and those on land--an important aspect of the proposed program which had not been satisfactorily resolved in the first. This aspect of research should include detailed comparative studies of fossil features on land which are recognized in the marine sections by geophysics, such as the accretionary wedge and back- and inter-arc basins. Thus, more detailed field work on land is needed.

Also, attempts should be made to unravel the geology of those complex areas in which platelets have impinged on each other, forming complicated patterns and terranes which are as yet little understood. These studies might constitute new or component parts of ongoing IGCP projects. Some more specific ideas about this subject are included in Dr. Katili's paper.

2) A link should be encouraged between pure geological research and its practical application; it is illustrated by an example of urban geology (also an aspect of Quaternary geology), that is, the sinking of cities such as Manila, Jakarta, Shanghai, Tokyo and Bangkok (Nutalaya and Rau, 1981). Greater attention should be paid to expediting the application of new knowledge, methods, and techniques to these critical geological problems which directly involve millions of people.
Governments increasingly expect practical benefit from their support of geology, and they should be made aware of the results.

3) These considerations naturally led to the need for making geologic concepts more widely known and emphasizing their importance to society. Geologists need to popularize their science and help make planners aware of it.

4) Another type of unified study, involving "pure" geological investigative disciplines and applying them to the practical purpose of making resource assessments, is the so-called "basin analysis." Assessments made at various levels of mineral, hydrocarbon, and water-resource potential could be carried out for the basins in the Southeast Asian region; the method might be applicable either to individual basins or to parts of the region.

5) The initiation and management of large projects requires the participation of many of the organizations previously mentioned. Particular attention would have to be paid to management methods in order to secure effective collaboration and implementation. In all such projects, detailed mapping on suitable bases is essential; CGMW might play a part here.

6) Concerning geological education, it was agreed that IUGS should pay attention to the region's requirements by supporting educational programs, supplying lecturers, and advising on educational matters. Also, better education is the key to improved professional standards and the termination of dependence on foreign assistance. The intellectual resources of the region are not yet being properly utilized. Well organized workshops are an important and effective method of education and of imparting special skills.

7) It was thought that national geological societies in Southeast Asia were, with few exceptions, less active than they should be. These societies could play a positive part in promoting research, in bringing together academic, government, and company geologists, in inspiring young geologists, in representing the science, and in producing publications. IUGS is well placed to provide advice and assistance to such geological societies and could arrange

to send lecturers to them. A suggestion that a regional union of geological societies should be formed for mutual support, to facilitate meetings such as GEOSEA, and to promote publication, seemed an idea worth supporting.

8) It was agreed by the seminar participants that, at a more personal level, communication between geologists is vitally important and is to be encouraged and assisted. Geological newsletters, including those of regional circulation as well as cheaply produced local ones, are valuable and should be promoted. Perhaps a network of newsletters could be established.

In bringing the seminar to a close, the President of IUGS, Prof. Seibold, thanked the organizers for their efforts which had led to such a fruitful and successful meeting.

SUMMARY AND CONCLUSIONS

The results of the seminar, in terms of major conclusions and recommended actions, may be summarized as follows, taking into account the resume by Hepworth and the discussion summaries by Berger, Basset, and Kielan-Jawarowska:

1. Scientific research programs

a) Develop a "SEATAR II" coordinated land-sea research program along the lines of "SEATAR I," but with greater emphasis on resources studies and on rotation and accretion of terrains or platelets. This requires development of plans for the program through a meeting of experts, similar to the meeting held in Bangkok in 1973 to plan "SEATAR I." IUGS should assist CCOP and ESCAP to organize such a meeting, in cooperation with other appropriate international and local agencies.

b) Organize a coordinated program in geology applied to urban problems for cities such as Bangkok, Manila, Jakarta, Shanghai, and Tokyo. Such a program would involve workshops, research, and training, and could include seminars being planned by AGID in Kuala Lumpur, a possible workshop in Hong Kong, and a pilot project by the Department of Mineral Resources in southeast Thailand. The IUGS exploratory working group on human problems should consult with those interested in these and other related activities to help organize a coordinated regional program.

c) Develop a basin analysis program in Southeast Asia, including a map project to organize, compile, and evaluate goelogical and geophysical data related to the distribution of hydrocarbon and non-energy resources. Such a program will involve the coordinated efforts of existing CCOP and ESCAP projects in hydrocarbon assessment, heat-flow studies, Quaternary geology, geophysical data compilation, and stratigraphic correlation, as well as the operational framework of the Circum-Pacific Map Project and the work of CGMW. The program would be carried out jointly by CCOP and the Circum-Pacific Project, under the aegis of the IUGS Research Development Program, in consultation with CGMW. It would be an initial component of the Research Development Program's world-wide basin analysis project. Plans are being formulated jointly by IUGS, CCOP, CGMW, and CPCEMR.

2. Institutional development and management

a) Strengthen and maintain effective institutions and programs such as those in Southeast Asia or develop institutions and programs in regions that are less effectively served. Governments and international organizations need to have evidence of practical benefits and timely results. IUGS should offer guidance and assistance to the extent feasible in strengthening and orienting programs towards such benefits and results.

b) Where programs are developed or assisted by IUGS, efforts should be made to involve institutions and ongoing programs. To the extent that mapping is involved, guidance should be provided by CGMW.

c) Efforts should be made, with IUGS assistance, to strengthen local geological societies; if a regional geological society is organized in Southeast Asia, IUGS should assist to the extent feasible.

d) The success of CCOP seems to have been partly a result of the use of special advisors in its Technical Advisory Group. IUGS should call attention to this as a possible model for other regional organizations to consider.

3. Education and communication

a) Establish a newsletter for transmitting information on geological activities and research in the region (or possibly broaden an existing newsletter, such as that issued by CCOP/EA).

b) Circulate booklists of new publications and arrange a literature exchange mechanism.

c) Increase, coordinate, and regularize the holding of workshops and training seminars.

d) Develop a communication and information exchange network, possibly as a function of a regional geological society.

LIST OF SEMINAR PARTICIPANTS

A. S. Balasubramanian
  Asian Institute of Technology
  Bangkok, Thailand

M. G. Bassett
  (Secretary General, IUGS Commission
    on Stratigraphy)
  National Museum of Wales
  Cardiff, United Kingdom

A. R. Berger
  (Editor, IUGS Episodes)
  Geological Survey of Canada
  Ottawa, Ontario, Canada

Zhang Bingxi
  Ministry of Geology and
    Mineral Resources
  Beijing, China

S. K. Chung
  Geological Survey of Malaysia
  Kuala Lumpur, Malaysia

F. Delany
  Commission for the Geological
    Map of the World
  Paris, France

E. Dubois
  Committee for Co-ordination of Joint
    Prospecting for Mineral Resources
    in Asian Offshore Areas (CCOP)
  Bangkok, Thailand

J. Fernandez
  Bureau of Mines and Geosciences
  Manila, The Philippines

H. M. S. Hartono
  Geological Research Development
    Centre
  Bandung, Indonesia

F. Hehuwat
  Institute of Geology and Mining
  Bandung, Indonesia

J. V. Hepworth
  ESCAP Regional Mineral Resources
    Development Centre
  Bandung, Indonesia

A. H. Hj Hassan
  Southeast Asia Tin Research and
    Development Centre
  Ipoh, Malaysia

J. H. Hodgson
  Regional Seismological Centre
    for Southeast Asia
  Manila, The Philippines

P. F. Howard
  (Vice President, IUGS)
  Macquarie University
  North Ryde, NSW, Australia

W. W. Hutchison
  (Chairman, IUGS Research
    Development Board)
  Dept. of Energy, Mines, and Resources
  Ottawa, Ontario, Canada

S. Jankovic
  United Nations Development Program
  Karachi, Pakistan

J. A. Katili
  Ministry of Mines and Energy
  Jakarta, Indonesia

G. Kautsky
  (Vice President, IUGS)
  Geological Survey of Sweden
  Uppsala, Sweden

C.S. Kenyon
  Cities Service East Asia, Inc.
  Singapore

T. T. Khoo
  University of Malaysia
  Kuala Lumpur, Malaysia

S. A. M. Kibria
    U.N. Economic and Social Commission
        for Asia and the Pacific (ESCAP)
    Bangkok, Thailand

Z. Kielan-Jaworoska
    (Vice President, IUGS)
    Academy of Sciences
    Warsaw, Poland

C. A. Kogbe
    International Centre for Training
        and Exchanges in the Geosciences (CIFEG)
    Paris, France

T. Liu
    Ministry of Geology and Mineral Resources
    Beijing, China

L. Machesky
    U.N. Economic and Social Commission
        for Asia and the Pacific (ESCAP)
    Bangkok, Thailand

A. Martinsson
    (Chairman, IUGS Commission on
        Stratigraphy)
    Uppsala University
    Uppsala, Sweden

W. H. Matthews III
    (Chairman, IUGS Commission on Geology
        Teaching)
    Lamar University
    Beaumont, Texas, USA

J. F. McDivitt
    ESCAP Regional Mineral Resources
        Development Center
    Bandung, Indonesia

V. V. Menner
    (Vice President, IUGS)
    Geological Institute
    Academy of Sciences
    Moscow, USSR

T. Nozawa
    (Circum-Pacific Map Project)
    Geological Survey of Japan
    Tokyo, Japan

C. Nishiwaki
    Institute of International Mineral
        Resources Development
    Shizuoka, Japan

V. Prakash
    UNESCO
    Jakarta, Indonesia

Prinya Nutalaya
    (President, AGID)
    Asian Institute of Technology
    Bandung, Indonesia

J. A. Reinemund
    (Treasurer, IUGS)
    U.S. Geological Survey
    Reston, Virginia, USA

E. Seibold
    (President, IUGS)
    Deutsche Forschungsgemeinschaft
    Bonn, Federal Republic of Germany

Sanarm Soensilpong
    Department of Mineral Resources
    Bangkok, Thailand

B.K. Tan
    University of Malaysia
    Kuala Lumpur, Malaysia

J. M. Tater
    Department of Mining and Geology
    Ministry of Industry and Commerce
    Kathmandu, Nepal

R. I. Volkov
    USSR National Committee of Geologists
    Moscow, USSR

E. von Braun
    (Secretary General, IGCP)
    UNESCO
    Paris, France

D. Wiercimok
    (Admin. Assistant, IUGS)
    Deutsche Forschungsgemeinschaft
    Bonn, Federal Republic of Germany

PART II: TECTONIC FRAMEWORK, RESOURCES OCCURRENCES AND
RELATED PROBLEMS IN SOUTHEAST ASIA

by

John A. Katili

Department of Mines and Energy

Jakarta, Indonesia

INTRODUCTION

The International Union of Geological Sciences asked the author to present a keynote address at the IUGS Seminar on Development and Coordination of Regional Geological Activities in Developing Countries, held in Bangkok in January 1983, regarding the geologic and resources framework and problems in the Southeast Asia region.

This is indeed a difficult task to perform because we are dealing here not only with the geology of several countries whose rocks range in age from Precambrian to Quaternary, but also because of the complex position of Southeast Asia in an area of interaction between three gigantic crustal blocks--the Eurasian, Indian Ocean, and Pacific Plates.

In the past, Southeast Asia has been well known as a testing area of earth science hypotheses. In particular, the theories put forward by Wegener (1922), Smith Sibinga (1933), and Kraus (1951) to account for the geology of this region were ahead of their time, and although they can no longer be fully accepted, such ideas as continental drift and undercurrent processes (now termed subduction) nevertheless constitute a part of the fundamental basis of the new global or plate tectonics. It was here that the first marine geophysical investigations were carried out about half a century ago, leading to the discovery that negative gravity anomalies coincide with the location of deep sea trenches. It was also in this region that scientists established the fact that the depth of earthquake hypocenters increases from the trenches toward the continents.

The impact of the geotectonic theories of Vening Meinesz (1934), Umbgrove (1949), Kuenen (1935), and Van Bemmelen (1949) can be understood only if one realizes that these scientists were the pioneers who for the first time recognized the importance of integrating land and marine geology to understand the complex nature of Indonesian geology, an approach which several decades later led to the development of concepts known as the new global tectonics.

Within the last 10 years, the Southeast Asia Tectonics and Resources (SEATAR) programme, jointly sponsored by the Intergovernmental Oceanographic Commission (IOC) and CCOP, has spent about $30 million in this region for earth science research, and in 1976-1977 alone, no less than 10 research vessels from many advanced countries operated in the deep seas of Southeast Asia. Obviously, Southeast Asia is still an important region for testing new tectonic concepts!

Southeast Asia is also endowed with rich natural resources such as oil and gas, tin, nickel, copper, and bauxite; hundreds of millions of dollars have been spent in exploring for these mineral riches over the last 20 years. At the moment the oil companies are spending more than one billion dollars annually for research and development in this region. Some of the notable spinoffs from oil exploration are the development of palynology as a valuable tool in stratigraphy, the determination of isotopic ages of a large number of rocks, and regional investigations of basinal framework, structural phenomena, and environmental conditions, as a result of which the shallow seas around Southeast Asia have been studied in as much detail as the North Sea.

It is therefore not surprising that currently in this classical area, cross fertilization of scientific concepts is intensively taking place between the mineral and petroleum industries, national institutions and many renowned international scientific bodies; it is to be expected that this process will result in the development of sound earth science theories which will be used in the discovery of new, and as yet unknown mineral and hydrocarbon deposits.

Some of the glimpses of new scientific discoveries and their bearing on mineral and hydrocarbon exploration in this region are discussed in this presentation.

BROAD TECTONIC FRAMEWORK OF SOUTHEAST ASIA

Cenozoic tectonic framework

The late Cenozoic crustal structure of the Southeast Asian region is characterized by the interaction of at least three major lithospheric plates. The plate boundaries of the region exhibit a variety of characteristics (fig. 1).

The boundary between the Indian Ocean and Eurasian Plates west of Sumatra and south of Java and the Lesser Sunda islands is a covergence-type boundary characterized by arc-trench systems, complicated in Sumatra by the presence of the 1350-km Sumatran Fault system. The Sumatra arc-trench system was formed solely by subduction of oceanic crust under continental crust. The continental crust is thick and relatively old here as it comprises volcanic-plutonic arcs of Late Carboniferous-Early Permian, Permian, Early Triassic, Cretaceous-early Tertiary and mid-Tertiary ages. The magmatic rocks formed above the Benioff zone are mostly silicic and intermediate in character. The down-going slab west of Sumatra dips at an angle of 5^0 beneath the inner slope of the trench, and bends somewhere between the outer-arc ridge and the Sumatra coastline to dip at 45^0 from a depth of 75 km to at least 183 km. At least two tears exist in the down-going plate, and these may in part be due to the oblique nature of the subduction processes (Kieckhefer, 1980). The Sumatra fault system is basically of dextral origin (Katili and Hehuwat, 1967), and movements of several hundreds of kilometers are indicated by the presence of long slices of oceanic crust

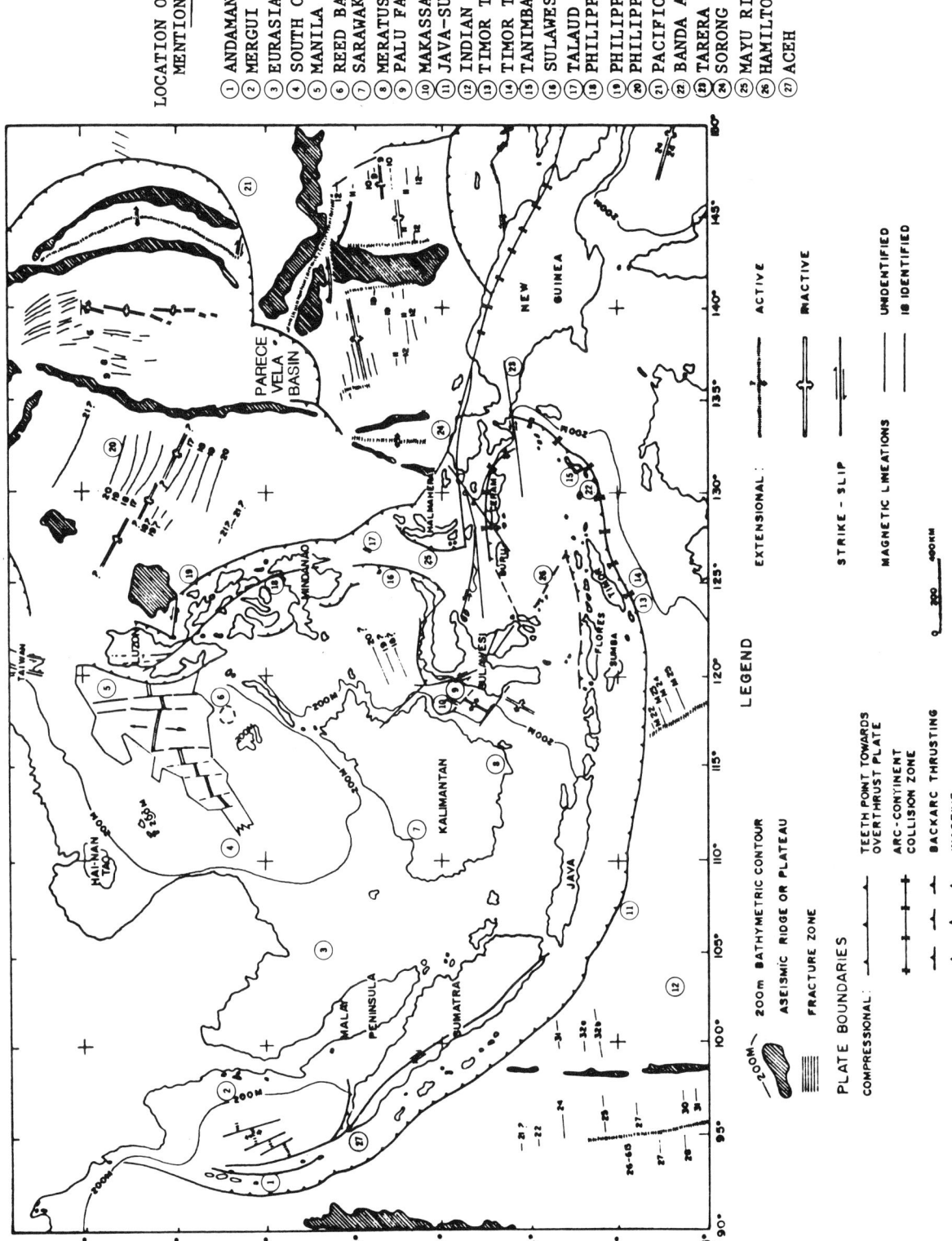

Figure 1. Present and Cenozoic tectonics of Southeast Asia (Modified from CCOP-IOC, 1980).

within the fault system, paleomagnetic evidence, and by the juxtaposition of two different geochemical provinces (Page and others, 1979).

A model for arc-trench slope sedimentation was developed by Moore and others (1980) in the Nias beds off the west coast of Central Sumatra. Sediments accumulated in small basins near the base of the slope by slumping from topographic highs and hemipelagic settling. Higher on the slope, some thrusts have become inactive, allowing the basins to combine. Terrigenous detritus from the arc is fed into the basins by canyons.

The Java arc-trench system was also formed by subduction of oceanic crust under continental crust. The continental crust is thin here, as it consists mostly of volcanic-plutonic arcs of Tertiary age. Lateral variations in the composition of volcanic rocks, which systematically become more potassic in the direction of the hinterland, have been clearly demonstrated in the island of Java. The maximum depth of the Benioff zone is about 700 km.

Eastward from Timor the trench shows an entirely different character in comparison with the Java-Sumatra Trench. Two distinct phases can be discerned in the development of the Banda arc. Initially only oceanic crust of the Indian Ocean Plate was subducted under the Banda oceanic plate, but as Australia drifted northward, continental crust entered the trench. The magmatic rocks formed above the Benioff zone are intermediate and mafic. The crust beneath the arc is thin and young, and is flanked on both sides by oceanic crust.

Most geologists maintain that the Banda Sea is underlain by a single tightly curved plate wrapped around the Banda arc from Timor to Ceram, although Cardwell and Isacks (1978) distinguished two separate Benioff zones, one dipping northward beneath Timor and the other separated by the Tarera-Aiduna transform fault, southward beneath Ceram.

For the most part the Philippines appear to form a separate lithospheric block between two trenches. The Philippine Trench to the east of Luzon and Mindanao has recently been more active and is associated with a westerly dipping Benioff zone resulting from subduction of the Philippine Plate. Activity of the Philippine Trench extends as far north as the island of Luzon in the form of thrusting. No accretionary wedge is associated with this trench. To the west of Luzon lies the Manila Trench, along which subduction of the South China Sea has occurred. The active north-south Manila Trench defines the eastern margin of the South China Sea. Mantle earthquakes indicate the presence of an active Bennioff zone. Volcanic rocks erupted above this zone range in composition from tholeiitic in the west, through calc-alkaline in the middle, to high potassic in the east. Between the two opposing subduction zones runs the 1200-km, sinistral transcurrent Philippine fault. Its relation to the tectonic features described above is not yet clear.

The Philippines are now beginning to be regarded as a collage of fragments: pieces of older continents or continental margins, oceanic sheets or ophiolites, and abundant components of arc systems, which act as matrix between the other terrains (Karig, 1981).

The 2500-km long Sorong transform fault system is considered to be the boundary between the Pacific and the Indian Ocean Plates.

The Eurasian, Indian Ocean, and Pacific Plates converge to a complex plate junction in the eastern part of Indonesia. The continuous northward movement of the Indian Ocean Plate and the westward thrust of the Pacific Plate along the Sorong transform fault, accompanied by the counter clockwise movement of New Guinea, have produced complicated tectonic structures such as the K-shaped form of Sulawesi and Halmahera (Katili, 1978), reverse polarities of small subduction zones around the northern arm of Sulawesi, active collision between the Sulawesi-Sangihe arc and the Halmahera arc (Silver and Moore, 1981), the formation of the 1000-km long sinistral Palu-Koro transform fault, and the opening of the Makassar Strait following minor sea floor spreading between Kalimantan and Sulawesi in Pliocene-Quaternary times.

Major elements of the mid-Tertiary tectonic framework are depicted in fig. 2. An attempt has been made (Katili, 1975) to restore the Sunda arc and the Sulawesi-Mindanao arc to their configuration prior to collision of the Eurasian, Indian Ocean, and Pacific Plates. It is envisaged that during that time the Sunda arc-trench system stretched from the northwestern tip of Sumatra through Java, the Lesser Sunda Islands, Timor, Tanimbar, and Kai to Ceram, exhibiting an east-west trend, with the Benioff zone dipping at a relatively steep angle towards the Asian continent. Intensive volcanism accompanied the subduction, the products of which are now well exposed along the west coast of Sumatra, the south coast of Java, and the Lesser Sunda Islands. Granitic rocks found in South Sumatra (Bengkunat granits), Java, Flores, Alor, and Ambon also belong to this volcanic-plutonic arc.

The Sulawesi-Mindanao arc, striking perpendicular to the Banda arc, is envisaged to exist about 600 km east of the present Sulawesi island. This volcanic-plutonic arc extends from western Sulawesi to Mindanao, whereas the Tertiary Sulawesi subduction zone follows the Talaud ridge, the submarine Mayu ridge and eastern Sulawesi. The extent of the subduction zone is confirmed by the presence of ophiolites or melanges in the islands mentioned, whereas the extent of the magmatic arc is confirmed by granites and volcanics that have been radiometrically dated (fig. 3).

It is postulated that the Tertiary Sunda-Banda arc had its origin in a spreading center situated in the Indian Ocean while the Sulawesi-Mindanao arc-trench system was generated by a spreading center located in the Pacific Ocean (Katili, 1974). The convergence history of the Banda and Sulawesi-Mindanao arcs is not clear but may have the character of a sinistral arc-to-arc transform fault. It is not unlikely that the Hamilton fault postulated by Silver (personal communication, 1983) constitutes the northwesterly extension of this several-hundred-kilometer-long Tertiary transform fault that linked the Sulawesi-Mindanao and the Banda arcs.

The boundary between the Pacific and Indian Ocean Plates which, as will be seen later, was not clear during the Mesozoic, is clearly defined by the east-west-trending Banda arc. Back-arc spreading within the Eurasian Plate occurred at this time. A northeast-southwest spreading axis has been

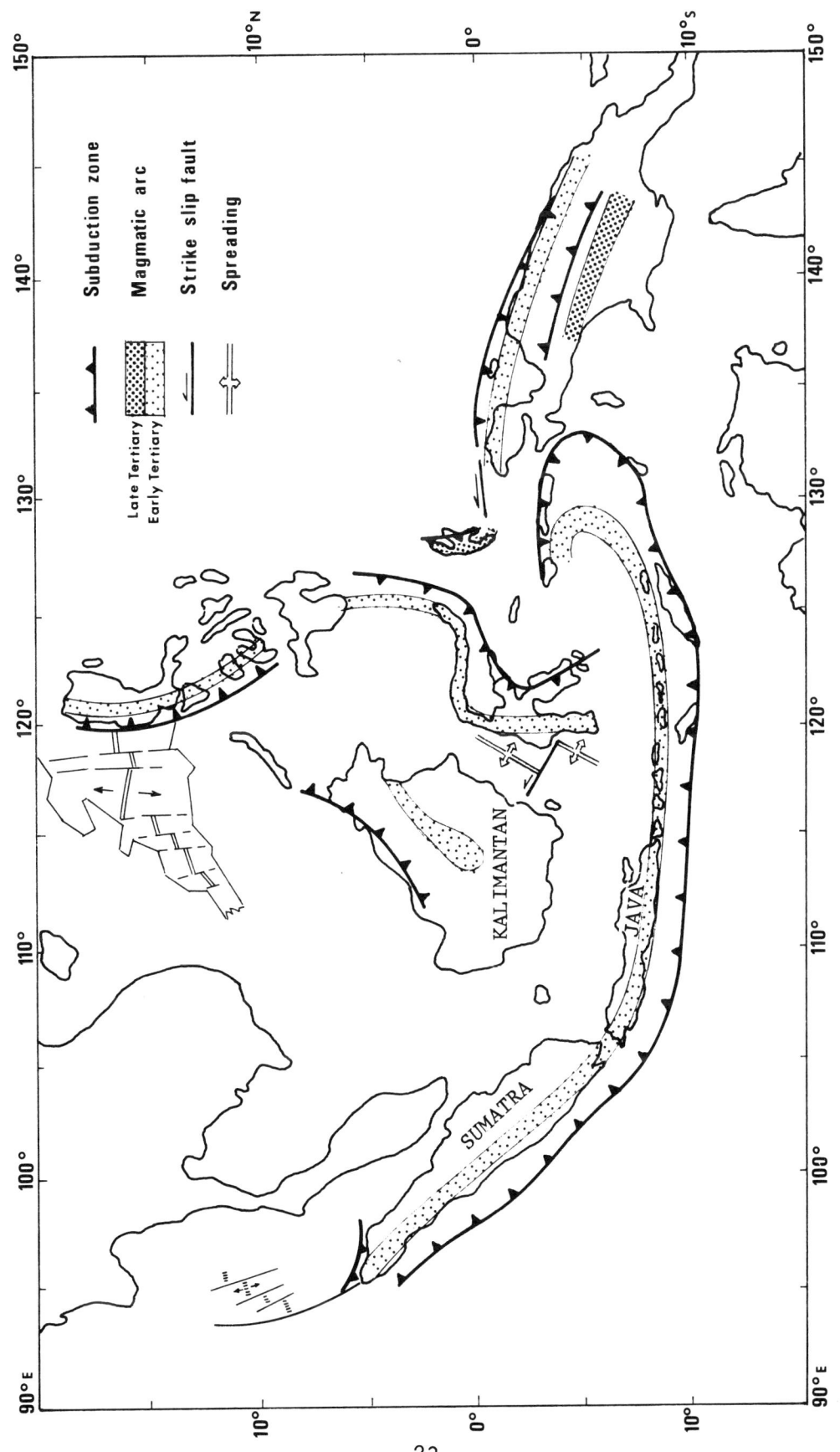

Figure 2. Tertiary tectonic framework of Southeast Asia.

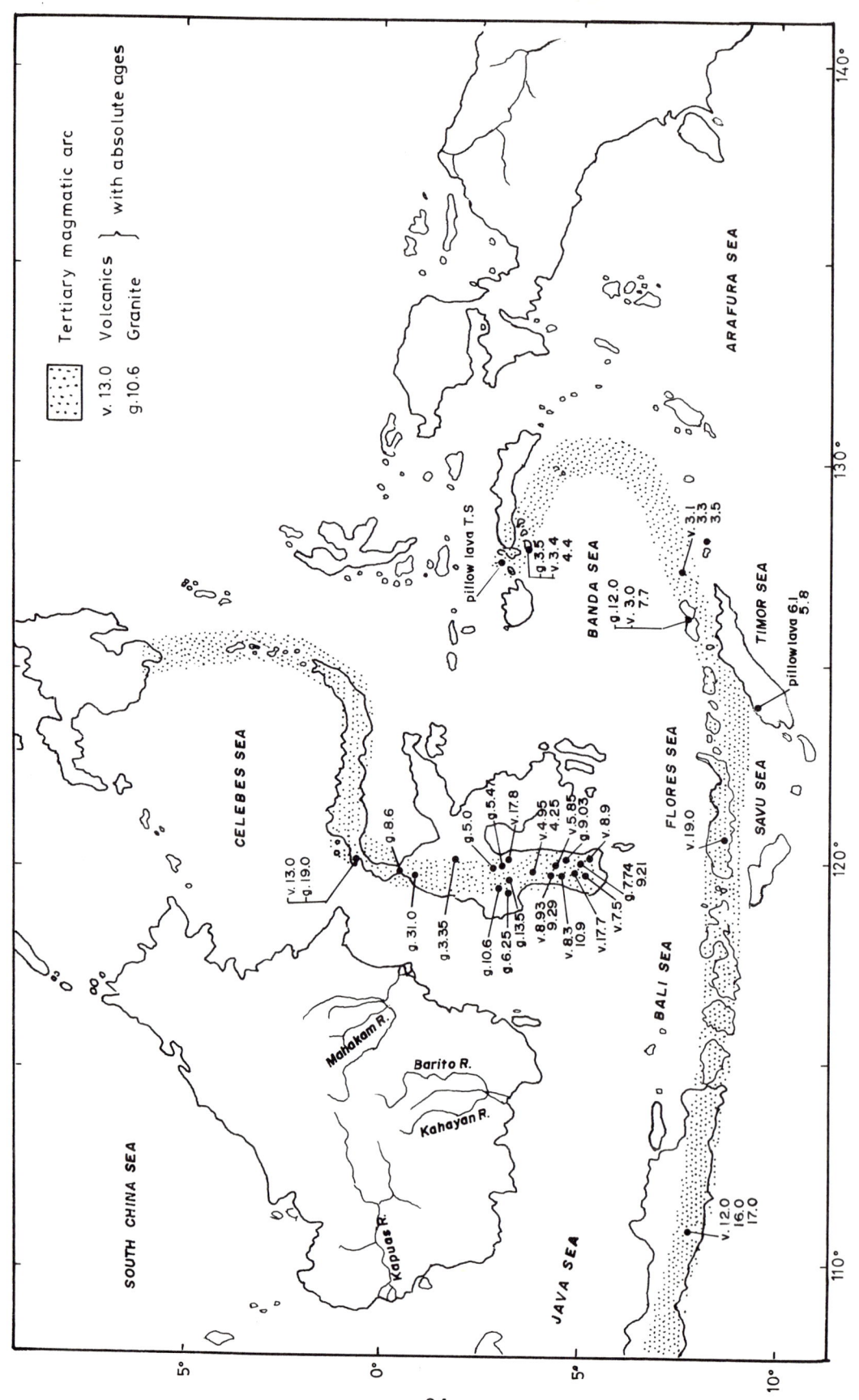

Figure 3. Radiometric data supporting the existence of the Tertiary Sunda-Banda arc.

identified in the Andaman Sea (Curray and others, 1979), characterized by high heat flow values and divided into several short segments by transform faults. The Andaman Sea spreading axis has developed since mid-Miocene time, pushing the Andaman-Nicobar ridge westward across the Indian Ocean floor. The Mergui Shelf of the eastern side of the Andaman Sea resembles an Atlantic-type continental margin produced as a result of rifting.

In the South China Sea region recent exploratory drilling on the Reed Bank has supported the previous concept that the intermediate shelfal region lying at about 2000 m represents the fragmented remains of continental crust. Oceanic crust emplaced between 32 and 17 m.y. (Taylor and Hayes, 1980) is identified as occupying the abyssal plain (fig. 4). Magnetic patterns of the basin give an eastward alignment of the spreading axis. The subduction zone which was formed by the spreading of the South China Sea is preserved on Palawan island and northwest Borneo. The Neogene volcanics in North Kalimantan are also associated with this subduction. Prior to this event, southerly directed Cretaceous and Paleogene subduction was active beneath northern Sundaland, which existed as an accreted crystalline core dating from the Palaeozoic.

Mesozoic tectonic framework

The Mesozoic tectonic framework of Southeast Asia is relatively simple, as the generation of arc-trench systems in the region was confined to spreading centers situated in the Indian Ocean and South China Sea. It is envisaged that during the Mesozoic, the eastern part of Indonesia was part of the proto-Indian Ocean and consisted of several microcontinents.

No Mesozoic island arc is known in this region and attempts to relate the scarce pre-Tertiary deposits in Sulawesi to a Mesozoic arc-trench system have failed so far. Presumably, the best way to explain these pre-Tertiary deposits is by assuming that they are allochthonous terrains, pieces of older continental fragments or island arcs which were incorporated in the Tertiary arc-trench system. Sulawesi is a composite island arc (Katili, 1978) and is the product of the incorporation of foreign fragments into its geology, the abundant components of the Tertiary arc system acting as a matrix between the older terrain, a situation similar to that described by Karig (1981) in the Philippines.

The late Mesozoic tectonic framework in the western Indonesian region is depicted in fig. 5a. The Cretaceous-early Tertiary volcanic-plutonic arc present in Sumatra does not continue into Java but passes north of it, running parallel to the subduction zone situated in the southern and northeastern part of Java (Katili, 1971). A second arc-trench system with opposing subduction zone runs through Kalimantan, presumably generated by a minor spreading center located in the South China Sea. The Lupar ophiolite belt (Hutchison, 1973) could possibly be linked with the Meratus-Babaris subduction by a trench-trench fault (Cameron, 1981).

A Mesozoic back-arc basin similar to that of the Tertiary system might exist in South Sumatra, but a marginal basin caused by back-arc spreading as envisaged by Cameron (1981) is geometrically questionable as it would be situated too close to the volcanic arc.

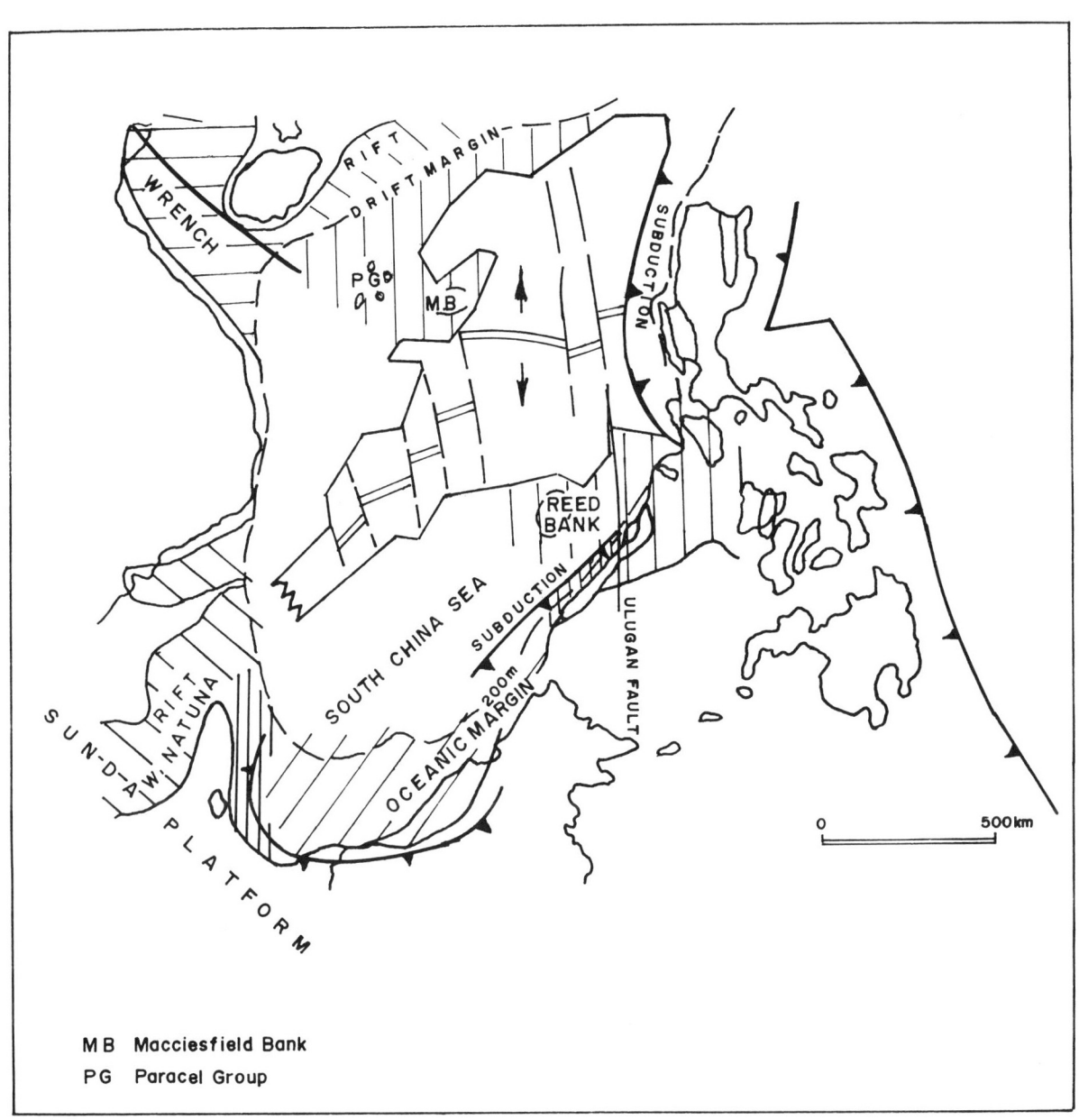

Figure 4. Tectonic style, South China Sea (Gage and Wing, 1980) (for explanation see text).

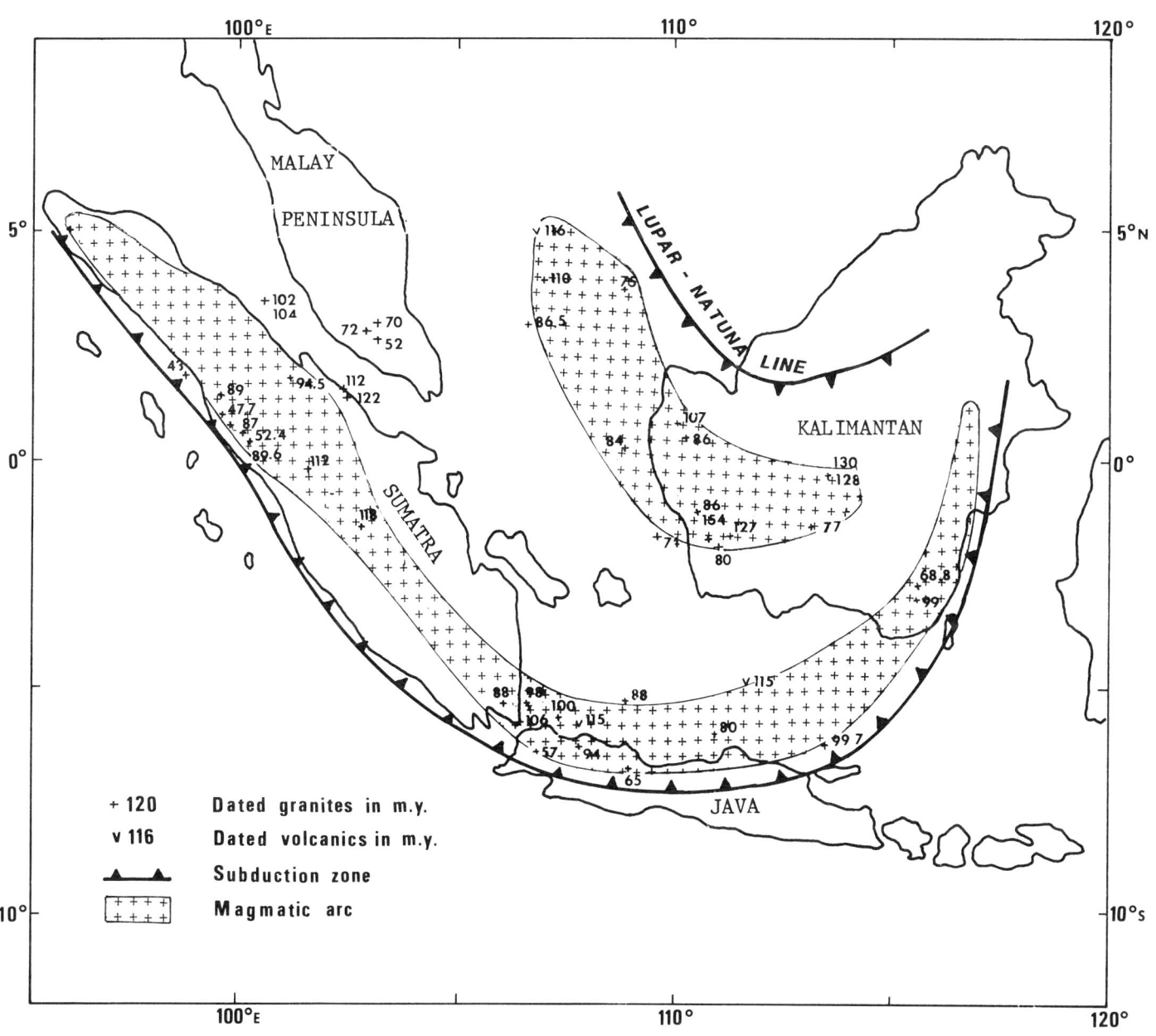

Figure 5a. Late Mesozoic tectonic framework of Southeast Asia.

The configuration of the Triassic-Jurassic island arc system was similar to that of the Cretaceous-early Tertiary one, demonstrating again the existence of a double island arc with opposing subduction zones (fig. 5b).

In comparison to the existence of the Cretaceous-early Tertiary arc trench system, which in general is accepted by most geologists, several investigators have rejected the concept of opposing Jurassic-Triassic island-arc systems. The plutonic arc that runs through the western part of the Malay peninsula and the Indonesian tin islands harbours the richest tin deposits in the world. Mitchell (1977) postulated that the Malaysian tin granites are related to collision of a West Malaysian plate with a volcanic arc lying to the east. The resulting suture is known as the Billiton-Bentong-Chiang Mai suture (Hutchison, 1975; Cameron, 1981; Suryono and Clarke, 1981). According to Cameron (1981), Sumatra, western Malaysia, and western Thailand can be considered as a sialic platelet that was attached to the remainder of Indochina along the Billiton-Bentong-Chiang Mai suture during the Late Triassic collisional orogenies. This implies a microcontinent to continent collision, although Cameron (1981) believes that a frequently active trench also lay off Sumatra from late Palaezoic time. He further stated that marine conditions persisted in north-western Sumatra during this collisional orogenesis, implying that the Thai-Malay orogeny may not be simply related to the Mesozoic trench of Sumatra.

Paleozoic tectonic framework

The presence of Permian volcanic and granitic rocks in Sumatra, in the Malay Peninsula and in western Kalimantan again point to the existence of two opposing Benioff zones in this region (fig. 5c). The Silungkang formation (Katili, 1969) in Central Sumatra characterizes a volcanic arc deposit, whereas Permian granites were encountered in drill holes in South Sumatra (Katili, 1973).

The second arc in West Malaysia was located east of the Bentong suture identified by Hutchison (1973).

An early Paleozoic tectonic framework similar to that of the late Paleozoic was reconstructed by Cameron (1981), who believed that a double arc-trench system with convergent Benioff zone existed in this region. The paleogeography of Sumatra during the early Paleozoic is not well understood. Regional high-grade metamorphics in Aceh might be related to the early Paleozoic tectonic framework (Cameron, 1981).

On the other hand, the geology of West Malaysia and Thailand is well understood. A volcanic arc existed in West Malaysia (Hutchison, 1973) during the Ordovician period. The volcanic-plutonic arc was flanked to the east by a fore-arc basin which give way westward to a stable shelf. A craton existed west of Thailand.

A basic model of two opposing subduction zones (Katili, 1973; 1974) in the western Indonesian region generated by a spreading center situated in the Indian Ocean and the South China Sea, operating throughout Phanerozoic time, may well explain the complicated geology of Sumatra, the Malay Peninsula, western Kalimantan, and the islands of the Sunda Shelf (fig. 6a).

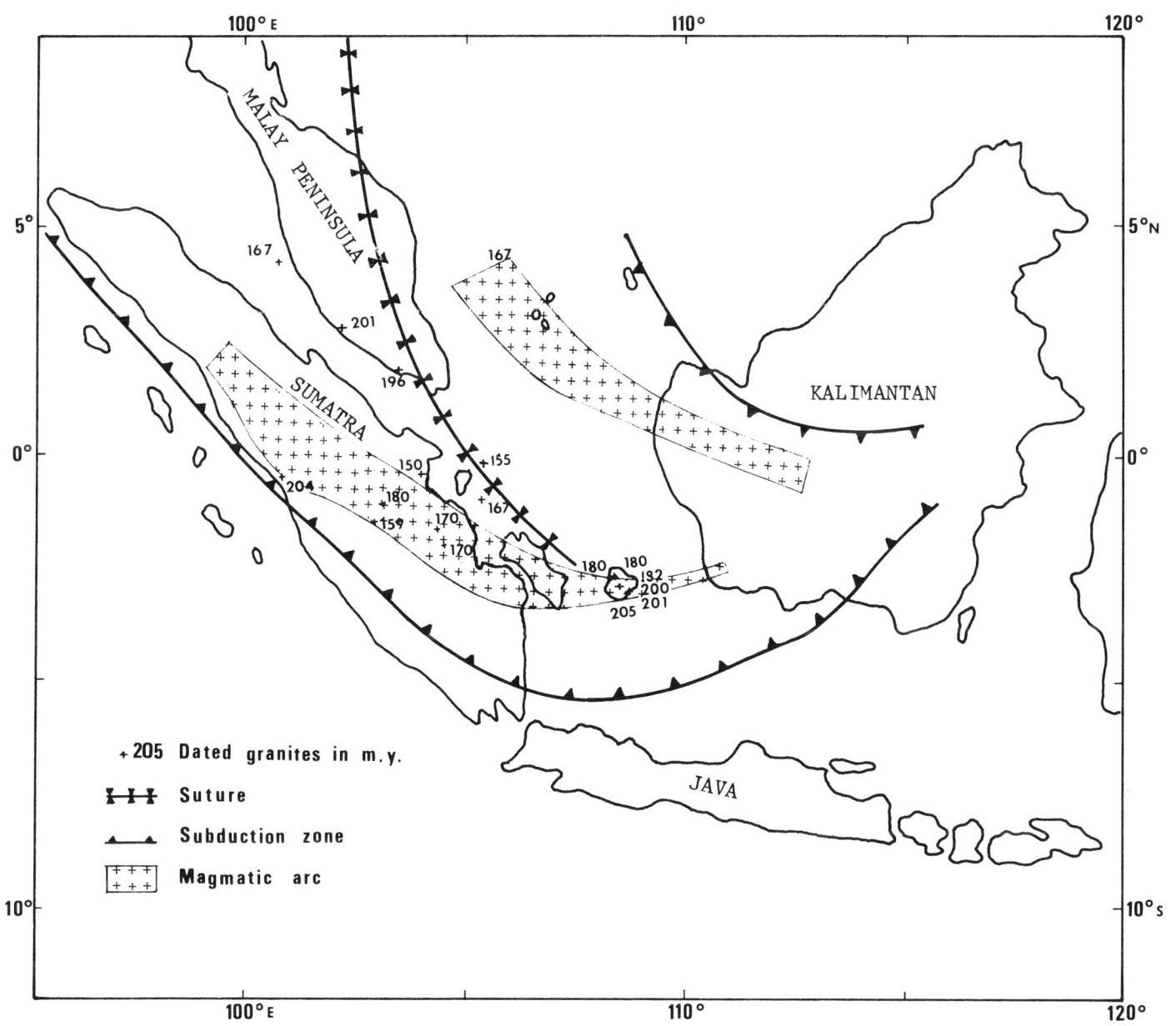

Figure 5b. Triassic-Jurassic tectonic framework of Southeast Asia.

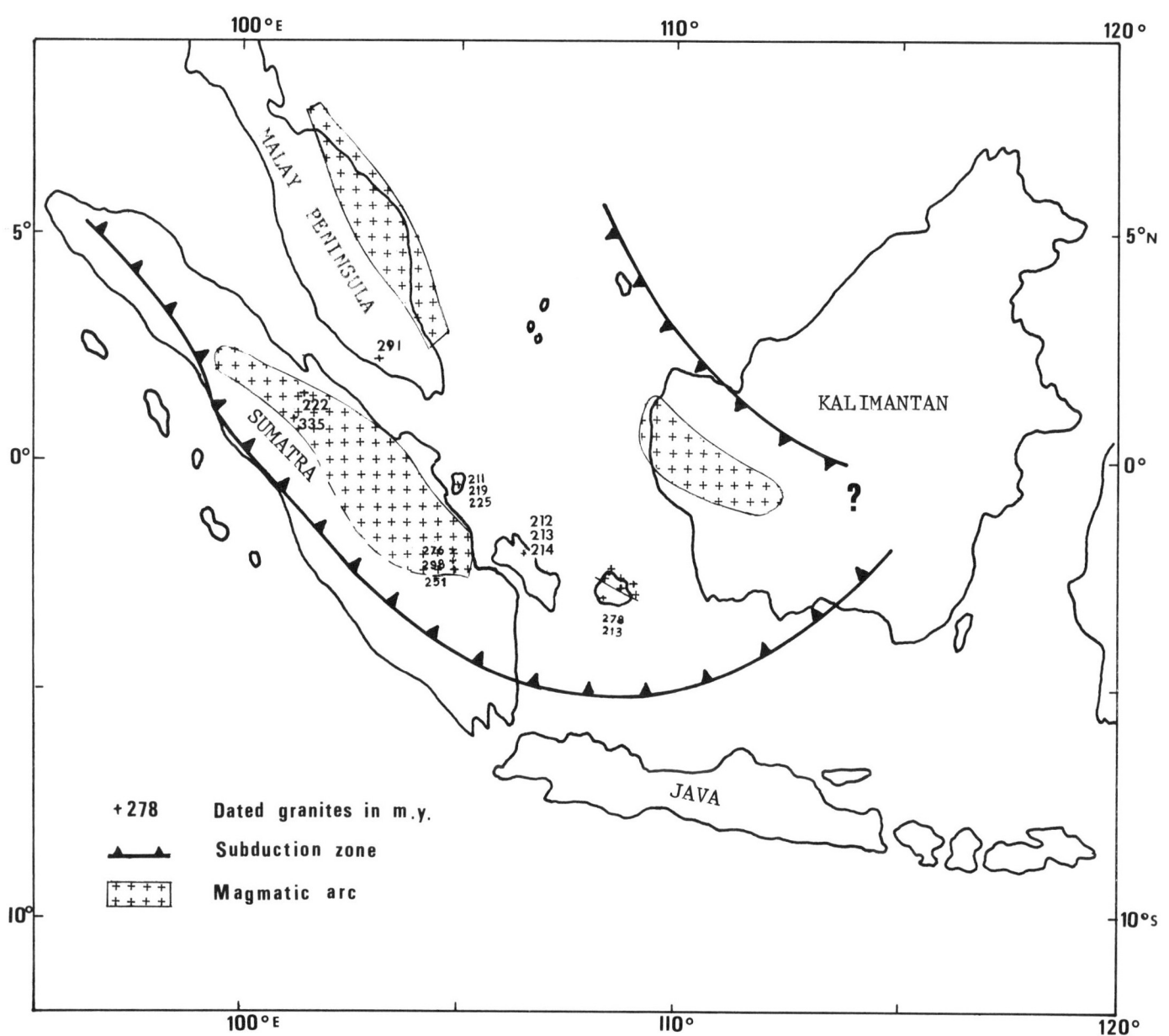

Figure 5c. Late Paleozoic tectonic framework of Southeast Asia.

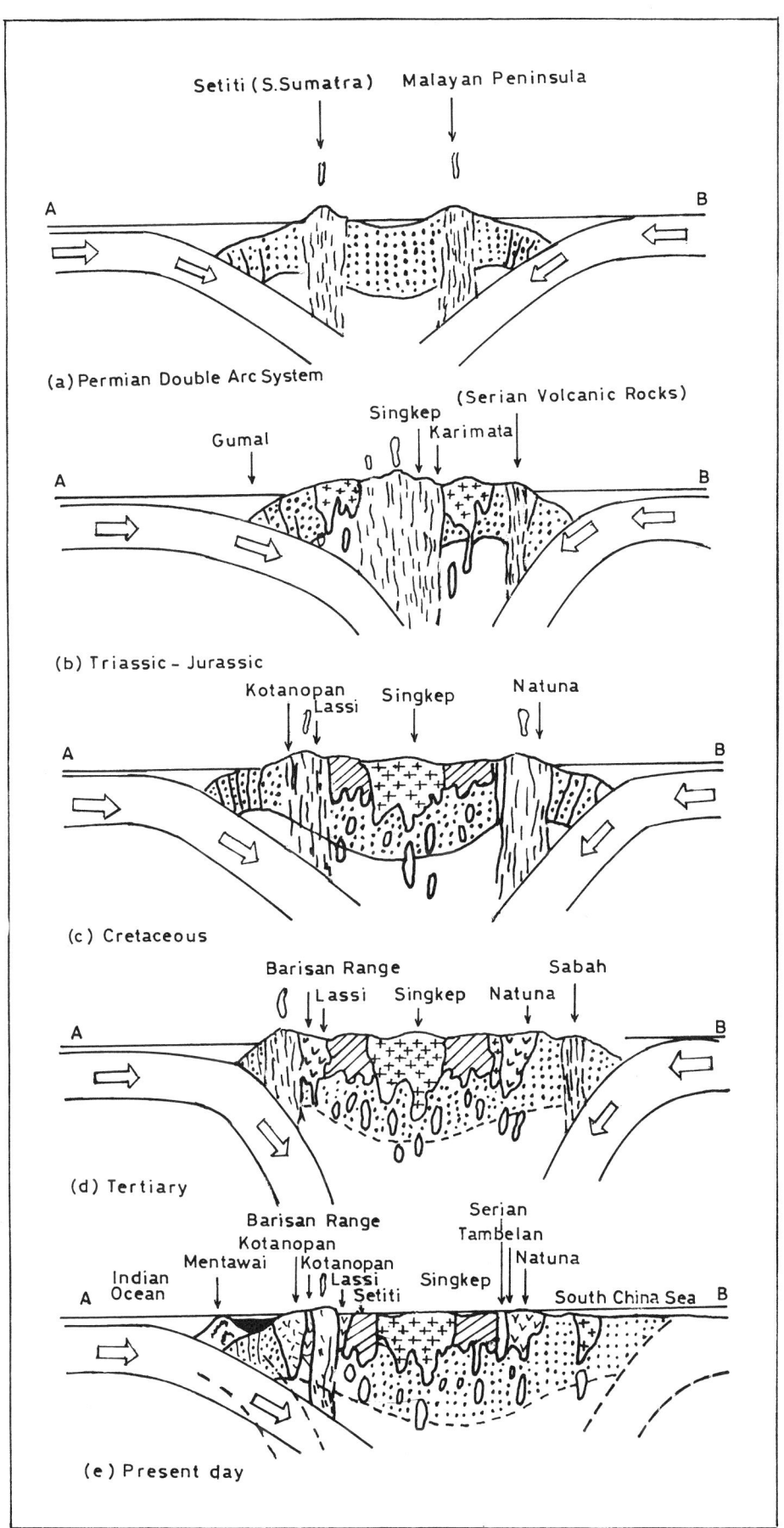

Figure 6a. Evolution of a double opposing arc-trench system of western Indonesia (Katili, 1973, 1974, 1980).

Modification to this basic model may become necessary as the detailed history of the complex mosaic of dominantly sialic, Paleozoic-cored platelets, linked by sutures (Hutchison, 1975), transcurrent faults (Katili, 1970), and extensional marginal basins (Curray et al, 1979; Rodolfo, 1969) becomes obvious. Fig. 6b shows the tectonic elements of western Indonesia and the South China Sea, including the various Phanerozoic basins as envisaged by Gage and Wing (1980).

Precambrian tectonic framework

Not much is known about the presence of Precambrian rocks in Southeast Asia. They occur with certainty only in Indochina (Hutchison, 1973). Their presence is suspected by Hutchison (1973) in West Malaysia, in Burma, and in Aceh by Cameron (1981).

It should be remembered that decades ago Stille and Lotze (1945) and Klompe (1957) had already postulated that the western part of Southeast Asia is the result of an outgrowth around the old Precambrian nucleus of Indochina.

TECTONIC EVOLUTION OF SOUTHEAST ASIA

In early Paleozoic times subduction occurred east of the Malayan Peninsula while the corresponding volcanic-plutonic arc came into existence in the central part of that area. A fore-arc developed to the east of the volcanic arc while to the west a back-arc came into existence. A subduction zone dipping towards the Asian continent is assumed to exist west of Sumatra. Thus since early Paleozoic time a double arc-trench system with opposing subduction zones existed in the western part of the Indonesian region.

In Late Carboniferous-Early Permian time, a subduction zone, dipping in the direction of the Asian continent, existed in or west of Sumatra. Andesitic volcanism and granitic emplacement in Sumatra accompanied this subduction process. Andesitic, basaltic, and granitic rocks encountered in the eastern part of the Malay Peninsula and western Borneo indicate that at the same time a subduction zone dipping toward the southwest may have existed at the northeastern margin of the continent. Subduction continued or shifted slightly oceanwards in Permian-Early Triassic time.

In Late Triassic-Jurassic time, the subduction zone at the southwestern continental margin shifted slightly again towards the Indian Ocean. The well developed, broad volcanic-plutonic arc of the Malay Peninsula and the Indonesian tin islands suggest that the Benioff zone was then probably shallower than the older subduction zones. Another minor subduction zone with opposing dip developed at the same time, presumably along the Lupar Line in Sarawak, indicating a slight migration of the northeastern subduction zone towards the South China Sea. The corresponding volcanic arc is represented by the Serian volcanic rocks and the Triassic volcanics encountered in drill holes in the Sunda Shelf. According to Mitchell (1977), a collisional event involving the two arc system took place in Triassic time, expanding the area of Sundaland and creating the tin granites of Malaysia, Banka, and Billiton.

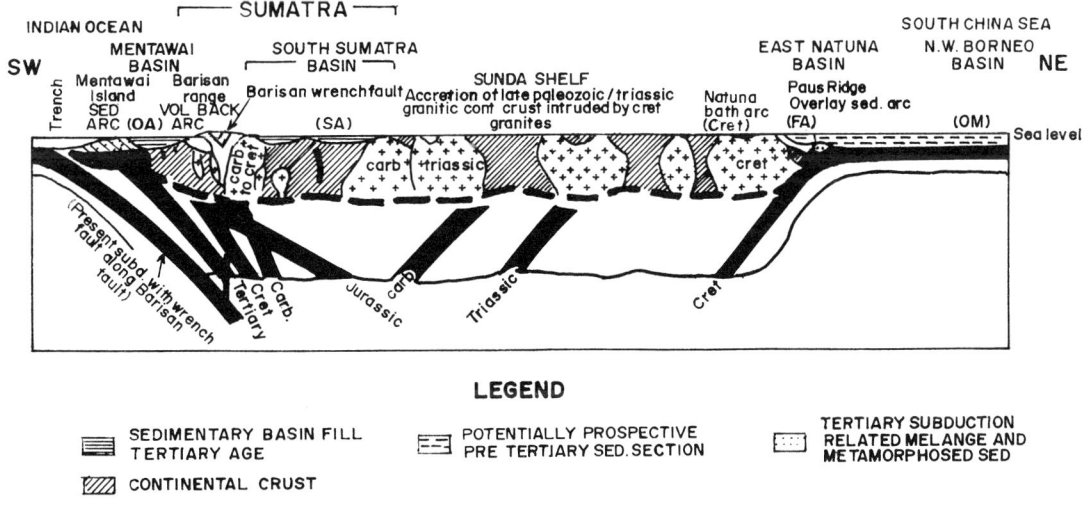

Figure 6b. Section across Mentawai, South Sumatra, Natuna Basin and South China Sea (Modified from Gage and Wing, 1980).

In the Early Cretaceous period a rift system within the Gondwana continent developed into a major spreading axis, resulting in the separation of India and Australia from the Gondwana block. Considerable significant reorganization of subduction systems took place, affecting Southeast Asia.

In Late Cretaceous-early Tertiary time, both the southwestern and northeastern subduction zones became larger as they moved towards the Indian Ocean and the South China Sea, respectively. Emergence of Sundaland occurred while subduction of oceanic crust continued under Borneo. Volcanic activity and emplacement of granites took place along the edge of the craton, including Natuna and the Anambas Islands.

During Tertiary time, the arc-trench system in Indonesia became fully developed. The new spreading center in the Indian Ocean generated an arc-trench system that stretched from the northwestern tip of Sumatra to Buru and Buton by way of Java, the Lesser Sunda Islands, Timor, Tanimbar, Kai and Ceram. The arc at that time comprised an approximately 6000 km-long Tertiary subduction zone dipping at a relatively steep angle towards the continent (fig. 7). Intensive volcanism was associated with the renewed subduction, and its products are now well exposed along the west coast of Sumatra, the south coast of Java, and the Lesser Sunda Islands. Granitic rocks in Sumatra, Java, Flores, Alor, and Ambon also belong to this Tertiary volcanic-plutonic arc.

Subduction along the Lupar-Natuna Line on the southwestern margin of the South China Sea ceased at the end of the Eocene. The Gulf of Thailand including the west Natuna sub-basin was formed by the collapse of the Sundaland mass.

In the Philippines, a westward-dipping subduction zone existed during late Eocene-early Oligocene time. This subduction generated the eastern belt of plutonic and volcanic rocks in the Philippines. This proto-Philippines arc collided with another arc situated on the downgoing plate and resulted in the cessation of the subduction process (Uyeda and McCabe, 1982).

In the Oligocene, andesitic volcanism started again in Sumatra and lasted until early Miocene time. It has been suggested that the mid-ocean ridge that migrated northward along the east side of the Ninety East Ridge collided with the western end of the "Old Sunda trench" in middle to late Micoene time (10-20 m.y. B.P). The ridge then jumped to the back arc region, and opened the Andaman Sea (Eguchi and others, 1979). Right-lateral strike slip movement along the Sumatra fault system was renewed during middle Miocene time. Emplacement of serpentinized late Mesozoic ultramafic rock occurred at the same time in northern Sumatra.

In the western part of the Philippines, subduction commenced in the Oligocene. This subduction brought the Philippines westward relative to the Eurasian Plate. The Parece Vella basin and Shikoku basin located on the eastern side of the Philippines were opened during this period. This would eventually bring the Philippines in contact with the Palawan microcontinent. During the Miocene the Philippines, situated above the eastward-dipping Benioff zone, collided with the Palawan block, resulting in the slowing down or cessation

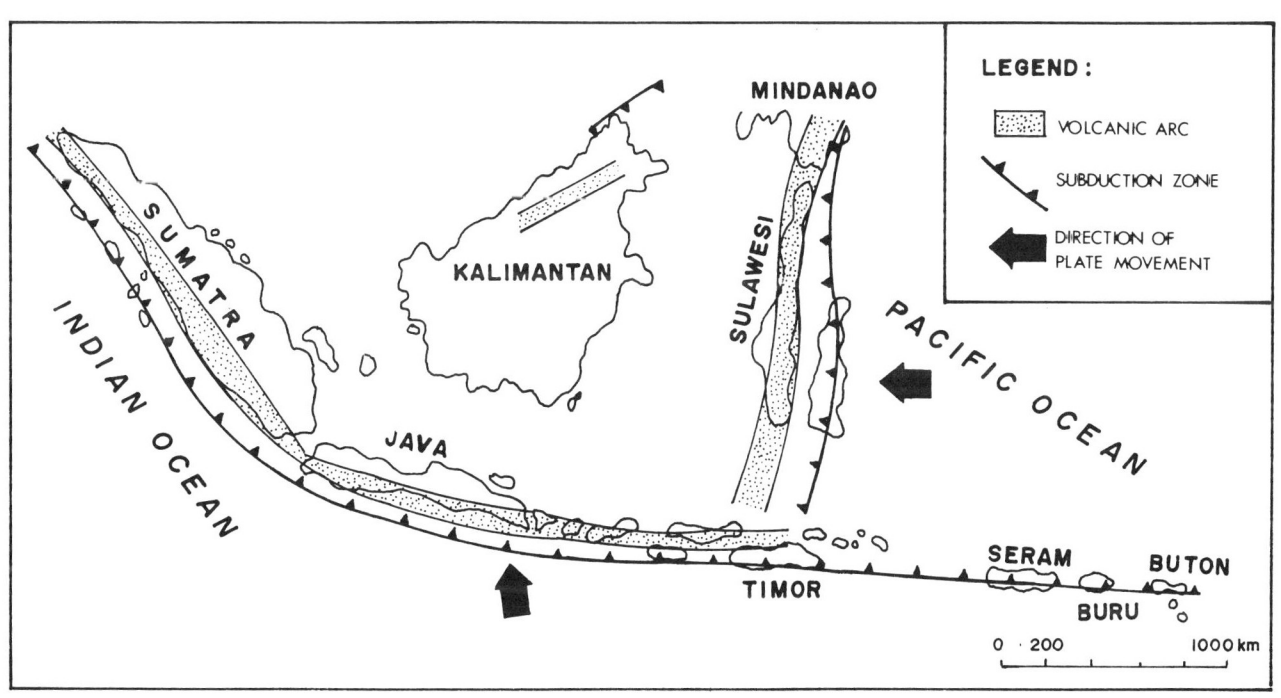

Figure 7. Evolution of the Indonesian island arcs in early Tertiary (Katili, 1974).

of subduction along the western boundary of the Philippine Sea Plate. Spreading within the Parece Vella basin ceased during this period (Mrozowski and Hayes, 1979). A flip of subduction occurred and the subduction zone was brought back to the eastern side of the Philippines (Uyeda and McCabe, 1982), (fig. 8).

In the South China Sea, new oceanic crust was emplaced resulting in the separation of the Reed Bank (fig. 1) from the Macclesfield Bank. The seafloor spreading which lasted until the early Miocene caused the subduction of the southward-advancing crustal block beneath the Borneo-Palawan ridge. The northern Palawan microcontinent drifted further south along the Ulugan transform fault, resulting in the attachment of this microcontinent to the south Palawan subduction zone. The collision of the Reed Bank with south Palawan probably resulted in the emplacement of ophiolites on Palawan during the Miocene.

In Miocene time or perhaps even earlier, a north-trending, east-facing island arc began to form 600 km east of Borneo, originating from a spreading center situated in the Pacific Ocean. This island arc marked a new pattern of subduction. The emergence of the Sulawesi-Mindanao island-arc system coincided with the change in directional movement of the Pacific Plate to west-northwest in Eocene/Oligocene time.

In middle to late Miocene time this north-trending Sulawesi subduction zone migrated farther eastward and created the eastward-facing Halmahera island arc. This arc did not extend further south as its growth was hampered by the northward-advancing Australian continent with New Guinea attached to its northern border.

Subduction ceased, presumably at the end of the Miocene epoch, and the Indonesian nonvolcanic outer arc, Mentawai-Nias, Timor, Tanimbar, Kai, Buru, Ceram, and Buton, was uplifted. In Southwestern Palawan, subduction also ceased in late Miocene time.

In Pliocene time, the subduction zone west of Sumatra and south of Java shifted oceanward to the present Sumatra-Java Trench. However, late Cenozoic to Holocene volcanism migrated in the opposite direction, as the dip of the Benioff zone was much shallower than that of the mid-Tertiary zone.

The most dramatic event in the geologic history of Indonesia took place in the Pliocene when the northward-advancing Australian continent, coupled with the counterclockwise rotation of New Guinea, caused the westward bending of the east-trending Banda arc. The continuous westward thrust along the Sorong fault system severely modified the east-facing Sulawesi arc into a K-shaped pattern (fig. 9).

This collision caused obduction of the ultrabasic rocks of the eastern and southeastern arms and thrusting of these rocks over the island arc. Continuous westward-directed tectonic forces along the Sorong transform fault system and the Matano fault zone in Sulawesi gradually pushed Sulawesi towards the Asian continent against Kalimantan and thus closed the southern part of the ancient Sulawesi Sea at the end of the Pliocene. The closure of this sea resulted in the obduction of the Cretaceous-early Tertiary Meratus and Pulau

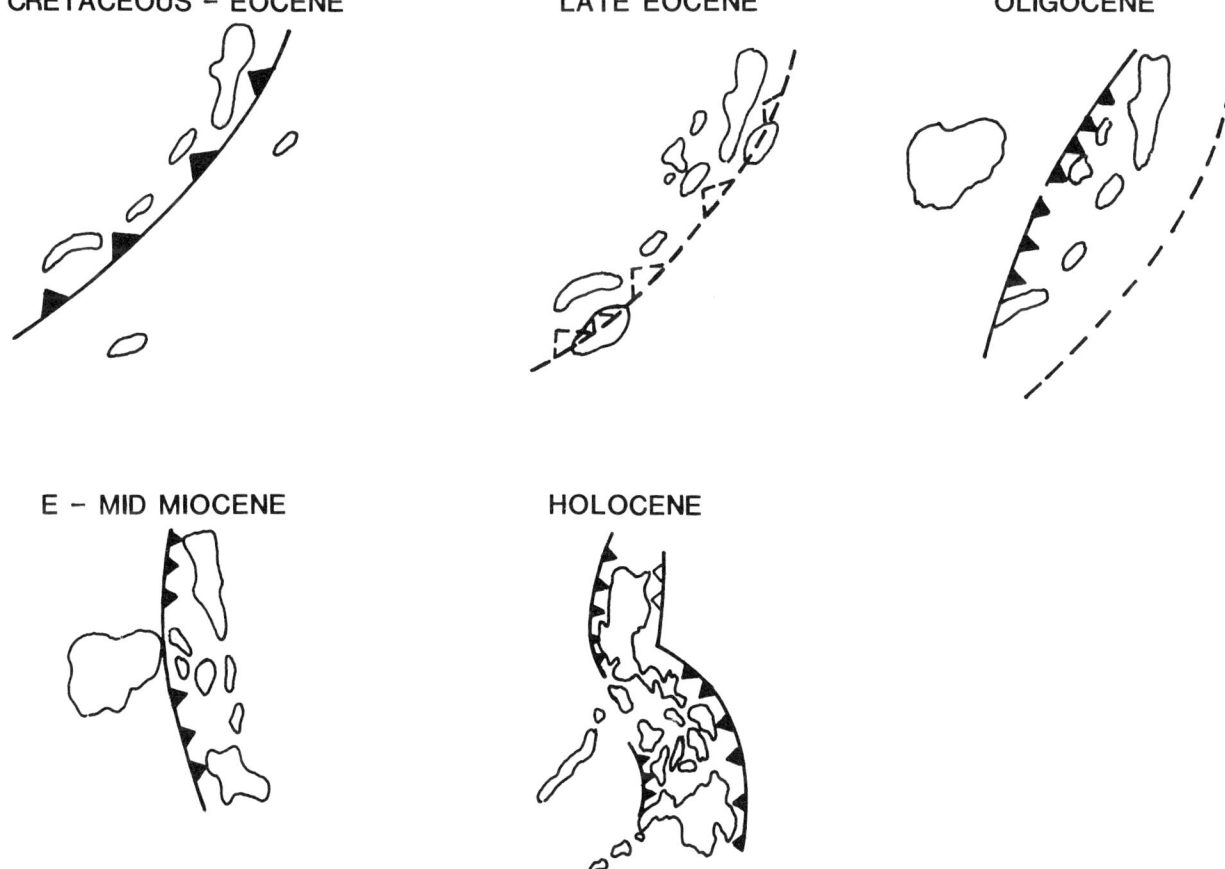

Figure 8. Possible stages in the Cenozoic evolution of the Philippines (Uyeda and McCabe, 1982)

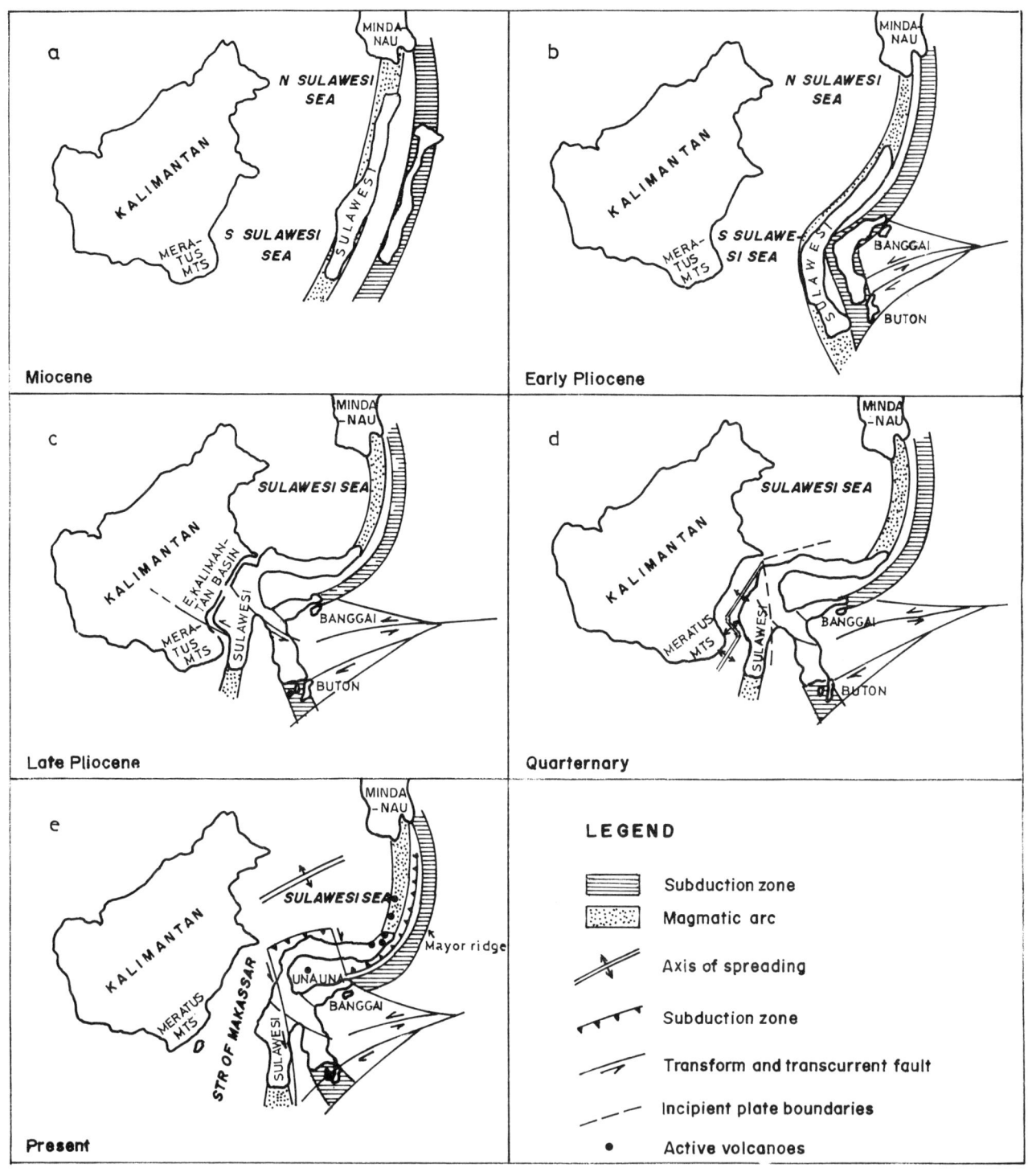

Figure 9. Geologic evolution of Sulawesi from Miocene till the Present Time (Katili, 1978).

Laut subduction complex and the formation of the Meratus mountain range. The rise of these mountains was not accompanied by plutonic activity; they were wholly caused by compressive forces, as there is no record of plutonic rocks of this age (van Bemmelen, 1949).

The South Sulawesi sea reopened, starting at the end of the Pliocene. The opening was caused by spreading along transform faults, the most important one being the Pater Noster fault (fig. 10). Eastward spreading south of the Pater Noster transform fault caused subduction and consequently created the now extinct Quaternary volcanoes of Lompobatang and Barupu in the southern arm of Sulawesi. The cessation of volcanicity was caused by subsequent spreading of the Sulawesi Sea which moved Sulawesi to the south-southeast along the Palu-Koro transform fault, simultaneously destroying the spreading centers in the Makassar Strait and thus cutting off the magmatic source of the Lompobatang and Barupu volcanoes. A small west-dipping subduction zone developed in northern Sulawesi accompanied by active volcanoes in Minahasa and the Sangihe islands. Other small subduction zones with reverse polarities subsequently developed in northwestern Sulawesi and Halmahera, which can be held responsible for the generation of the active Una-Una volcano in the Gulf of Gorontalo, central Celebes, and the volcanoes in the western Halmahera arc. Active collision accompanied by emplacement of ophiolites is presently taking place between the west-facing Halmahera arc and the east-facing Sangihe arc (Silver and Moore, 1981).

FUNDAMENTAL PROBLEMS OF AND CONTRADICTING VIEWS ON THE PLATE TECTONICS OF SOUTHEAST ASIA

Fundamental problems in active continental margins have been ably summarized in a report of the 3rd International Workshop on Marine Geosciences held in Heidelberg in 1982 (Theide, 1982).

Questions as to the origin of arc magmatism and the cause of back-arc spreading, the fate of the descending lithospheric slab, and horizontal and vertical movements in relation to tectonic stress remain unsolved. Recent studies have shown that in addition to complex structural variations, convergent margins also undergo frequent changes in character through time. Examples of these changes have been presented in the previous discussion on the tectonic evolution of eastern Indonesia and the Philippines. These variations include episodic vertical tectonics, back-arc spreading, arc rifting, and change in volcanic patterns in response to changes in global tectonic patterns.

Temporal variation must also be related to local effects, such as subduction of buoyant features and spreading centers.

Types of subduction

Uyeda (1982) distinguished two types of subduction zones, namely the Chilean type and the Mariana type. The Chilean type is characterized by the existence of a shallow trench, an accretionary prism, a fore-arc basin, calc-alkaline andesitic volcanism in the volcanic arc, porphyry copper deposits, absence of marginal basin, and lack of regional high heat flow. The Benioff

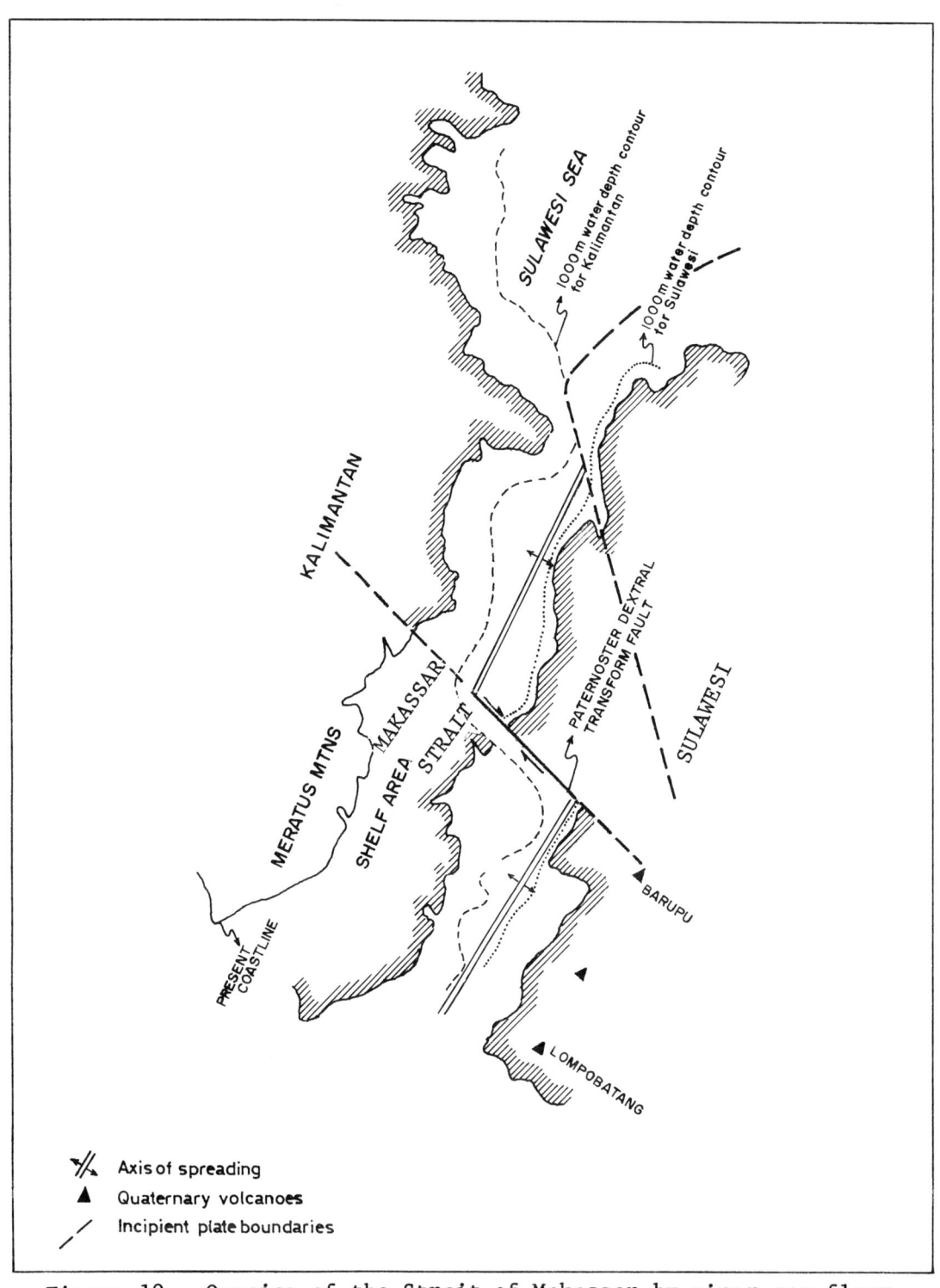

Figure 10. Opening of the Strait of Makassar by minor sea-floor spreading along the east-west-trending dextral Paternoster transform fault. Development of an east-dipping small subduction zone and subsequent late Quaternary volcanoes (Katili, 1978) in the southern arm.

zone exhibits a shallow dip, and great thrust earthquakes are common. Present-day stress in the volcanic arc area is compressional for the Chilean type arc. The Mariana type subduction is characterized by the existence of a deep trench, the absence of an accretionary prism, basaltrhyolite volcanism in the volcanic arc, rich Kuroko-type metalliferous deposits, marginal basins formed by tensional forces, and a high regional heat flow.

Uyeda (1982) considered the Sumatra arc as being of the Chilean type, but the existence of the marginal Andaman Sea at the northwestern end of this arc-trench system is not compatible with that theory. Moreover, as will be discussed later, no prophyry copper deposit of economic importance has been discovered in Sumatra so far, despite intensive exploration efforts (Katili, 1974).

Strongly curved arcs

Hamilton (1981) pointed out that the geometry of the tightly curved Banda arc, as well as the other strongly curved arcs around the world, is not compatible with the concept of the injection of a rigid plate. Such configurations are more easily explained by the sinking of a plastic plate. He went on to argue that reversals of polarity in zones of convergence as observed in many places in eastern Indonesia can be accounted for by the concept of a vertical sinking plate.

This concept of a sinking plastic plate reminds the present author of Van Bemmelen's paper outlining the present status of his undation theory. His geodynamic model accepts the continental drift and sea-floor-spreading hypotheses, but differs from plate tectonics in its mechanical concepts and its geochemical and geothermal view. It suggests a more rheological approach to structural evolution in which rigid plates represent only a surface skin effect (van Bemmelen, 1978). Based on this rheological view, van Bemmelen made a comparison between the Banda "whirl" of eastern Indonesia and another spiral structure which surrounds the southeast corner of the Darwin Rise. The present author has refrained from envisaging such strongly curved zones in western Indonesia but has instead advocated the occurrence of opposing subduction zones related to existing spreading centers. Such a tectonic framework does not challenge the validity of the concept of rigid plate subduction. The concept of a closed system of plate boundaries has perhaps led some authors to link the origin of the Banda arc to the origin of Sulawesi. The present author has made a distinction between the Banda arc, which is related to a spreading center in the Indian Ocean, and the Sulawesi arc, which was generated by another spreading center in the Pacific Ocean.

Seismicity and form of the Banda arc

Another controversial view centers around the seismicity and the form of the subducted Banda Plate. Most geologists maintain that the Banda Sea is underlain by a single, tightly curved plate surrounding the Banda arc from Timor to Ceram. Cardwell and Isacks (1978), using the most reliable seismic data, distinguished two separate Benioff zones, one extending northward beneath Timor to a depth of 600 km and the other, separated by the Tarera-Aiduna transform fault, extending southward beneath Ceram to a depth of about 200 km. This two-plate model could perhaps accommodate the loop-shaped arc without challenging the concept of rigid plate injection. However, Untung and Barlow (1981) as part of the joint geological and geophysical mapping programme with the Bureau of Mineral Resources (Canberra), could not discern the discontinuity in the Bouger anomaly pattern in the region of the Tarera-Aiduna fault which would justify a two-plate model for the Banda arc. Added to this is the fact that the nature of the eastern extension of the above-mentioned transform fault has not yet been identified. The two-plate model for the Banda Sea is also difficult to reconcile with the results of field investigations by Audley Charles and others, (1981), as their findings indicate that the stratigraphy and structure of Ceram show a remarkable resemblance to Timor. They argue that whatever hypothesis is used to describe the tectonic evolution of Timor would be equally applicable to Ceram, implying a preference for a one-plate model for this region.

The age of the Banda basin

The present author has advanced the hypothesis that the Banda Sea represents an old oceanic crust (Katili, 1974) trapped by the rolling up of the Banda arc. Bowin and others (1980) favoured a Cretaceous or older age for the Banda basin and considered that it is a trapped piece of oceanic crust. In contrast Hamilton (1977, 1979) and Carter and others (1976) suggest a late Tertiary age for part of the Banda Sea crust. They maintained that the Banda volcanic arc was originally continuous with the Sulawesi volcanic arc to the north, and associated with a west-dipping subduction zone. Late Tertiary back-arc spreading west of the Banda arc caused the eastward migration of this arc away from the Sulawesi arc, creating the Banda Sea. The curvature of the arc reflects the progressive collision of the arc with the curved continental margin of northwestern Australia and Irian Jaya (West New Guinea). Lapouille and others (in press) came to the conclusion that the Banda Sea crust had been formed in Early Cretaceous time as a part of the eastern Indian Ocean and western Pacific Ocean, and subsequently was trapped, most probably in Miocene time.

Rotation of Sulawesi

It has been stated earlier that the present K-shaped form of Sulawesi was the result of collision of the Sula Spur with the north-south trending Sulawesi arc. To test this hypothesis Sasajima and others (1980) has carried out palaeomagnetic studies on the north arm of Sulawesi. They proposed that the welding of the east and west arm of Sulawesi took place between 19-13 m.y.B.P., that is, between the Miocene and Pliocene, a conclusion accepted by the present author. They discovered that between the early to late Miocene, a rotation of nearly 40° had taken place in the northern arm. These results require either a counter-clockwise rotation, assuming the older rocks are normally magnetized, or a clockwise rotation, assuming reversed magnetization. Sasajima suggested a clockwise rotation about an axis near the neck of Sulawesi between the South and North arms, supporting the present author's hypothesis but on a slightly earlier time framework. Silver (personal communication) assumed that the hinge was in the east and that the north arm rotated clockwise along the Palu-Koro fault zone. This is in agreement with Hamilton's hypothesis of the back-arc spreading history west of the Banda arc.

Opening of the Makassar Strait

The contradicting views outlined above bring us to another problem pertaining to the age of the opening of the Makassar Strait. The present author suggested that the opening took place in Quaternary time. Other investigators, most notably Hamilton (1979), have proposed that Kalimantan and Sulawesi separated prior to the middle Miocene. If we agree with Sasajima that collision of the Sula Spur and Sulawesi occurred at the latest about 13 m.y.B.P. (middle Miocene), and assuming that Sulawesi was pushed 800 km westward at a rate of 10 cm/year, then it can be computed that the ancient Sulawesi Sea was closed around 5 m.y.B.P., that is, in Pliocene time. Subsequent separation must have occurred in post-Pliocene or Quaternary time. The Palu-Koro transform fault is active (Katili, 1970), and came into being concurrently with the generation of a new subduction zone with reversed polarity in the Celebes Sea. It is therefore likely that the rotation along this fault during Miocene time as suggested by Silver did not take place.

Position of Southeast Asia in relation to continental drift

There has been much debate as to whether Sumatra, the Malayan Peninsula, and Thailand formed part of Gondwanaland or Eurasia during the Mesozoic. Ridd (1971) is of the opinion that at least Thailand, the Malay Peninsula, Sumatra, Kalimantan, and Java formed part of Gondwana and were located between India and Australia until its breakup in the Mosozoic. On the other hand, McElhinny, Haile, and Crawford (1974), and Stauffer and Gobett (1972) maintained that the Malayan Peninsula was never part of Gondwana, but was situated at 15°N latitude in late Palaeozoic time. The present author has always supported the opinion put forward by McElhinny and others (1974) that the Malay Peninsula and Sumatra could not have formed part of Gondwana, as advocated by Ridd (1971), as it is obvious that the regular zonal structure of the Phanerozoic arc-trench system of the western part of Indonesia, as demonstrated previously, could only be explained by assuming that no significant drift had occurred in this region during the time span postulated above.

Rotation of Sumatra

The problem outlined above was once again raised by Sasajima and others (1980), who, using palaeomagnetic evidence, indicated that in the early Miocene the basement block of Sumatra was situated 10°-20° south of its present latitude. During the Mesozoic, part of Sumatra (a microcontinent?) drifted northward accompanied by a 40° clockwise rotation, reaching the present latitude in early Tertiary time. The occurrence of Glossopteris in northern Thailand may point to a Gondwana origin of some parts of Southeast Asia. If Sasajima is correct, then we have to assume that part of Gondwana (as microcontinents) has been incorporated in the Tertiary island arc system formed at the present site of Sumatra.

A clockwise rotation of 20° of Sumatra around an axis located in the Sunda Strait was also postulated by Ninkovich (1976), based mainly on the age difference of rhyolite flows in south and north Sumatra. Tjia (1977) however, suggested that this conclusion need not be valid, as the evidence for the position of North Sumatra was based on only one of the at least four separate ignimbrite eruptions that occurred in the Toba area.

In the petroliferous North Sumatra basin, Tjia and Najoan (1981) recognized at least four major zones of dextral slip faults that strike between north and N 10° W. Interruption and deformation of the various Tertiary seismic horizons indicate that this trend was inherited from the pre-Tertiary basement. Predominantly lateral slip motion has affected beds up to the level of the Lower Keutapang (upper Miocene-lower Pliocene). In the same basin the previously recognized Peusangan fault zone strikes in the same direction and has left-lateral motion. Similar major lineaments are also known from the basement of the Central Sumatra basin and from the pre-Tertiary rocks of Peninsula Malaysia (e.g. the Karak fault zone, Tjia 1972). If Sumatra has rotated during the Cenozoic, then this movement must have either affected Peninsula Malaysia (paleomagnetic evidence indicate no rotation), or that part of Sumatra west of the petroleum basins was fused to Sundaland during the Cenozoic. The Tertiary stratigraphy of this area, however, does not support this latter concept.

TECTONIC ENVIRONMENT OF THE SOUTHEAST ASIAN MINERAL AND HYDROCARBON DEPOSITS

Mineral deposits and tectonic framework

The tectonic framework as presented in the previous pages has been proposed to be used as a basis to construct a metallogenic map of Indonesia (Katili, 1974). The geology and mineralization of each "orogen" (Westterveld, 1952) can now be resolved by the plate-tectonic interpretation which links mineral provinces together in a more realistic way than was previously possible. An attempt was made by the present author (Katili, 1974) to distinguish arc-trench systems generated by spreading of the Indian Ocean and the ones which originated by spreading of the Pacific Ocean (fig. 11). From this exercise it was apparent that the Tertiary mineralization in Sulawesi, Halmahera, and New Guinea--all generated by spreading and subduction of the Pacific Ocean floor--is more significant in comparison with that of Sumatra, Java, and the Lesser Sunda Islands, which had their origin in the spreading center located in the Indian Ocean. This might be due to the fact that the crustal elements generated by the Pacific spreading center were, as evidenced by the occurrence of abundant polymetallic nodules in some parts of the Pacific Ocean floor, richer in metal than those of the Indian Ocean.

The abundant lateritic nickel deposits in Southeast Sulawesi, Gebe, Gag, and the Cyclop Mountains in western New Guinea (Irian Jaya) had their origin in the ultrabasic rocks found in the subduction zones related to the Pacific Ocean. These areas coincide with the Sulawesi-Moluccas collision zone and the Sorong transform fault where slabs of oceanic crust and mantle are thrust over and emplaced along the island arcs.

The distribution of the ores in the Tertiary volcanic rocks shows clearly that the metals are derived from oceanic materials, as the largest and richest

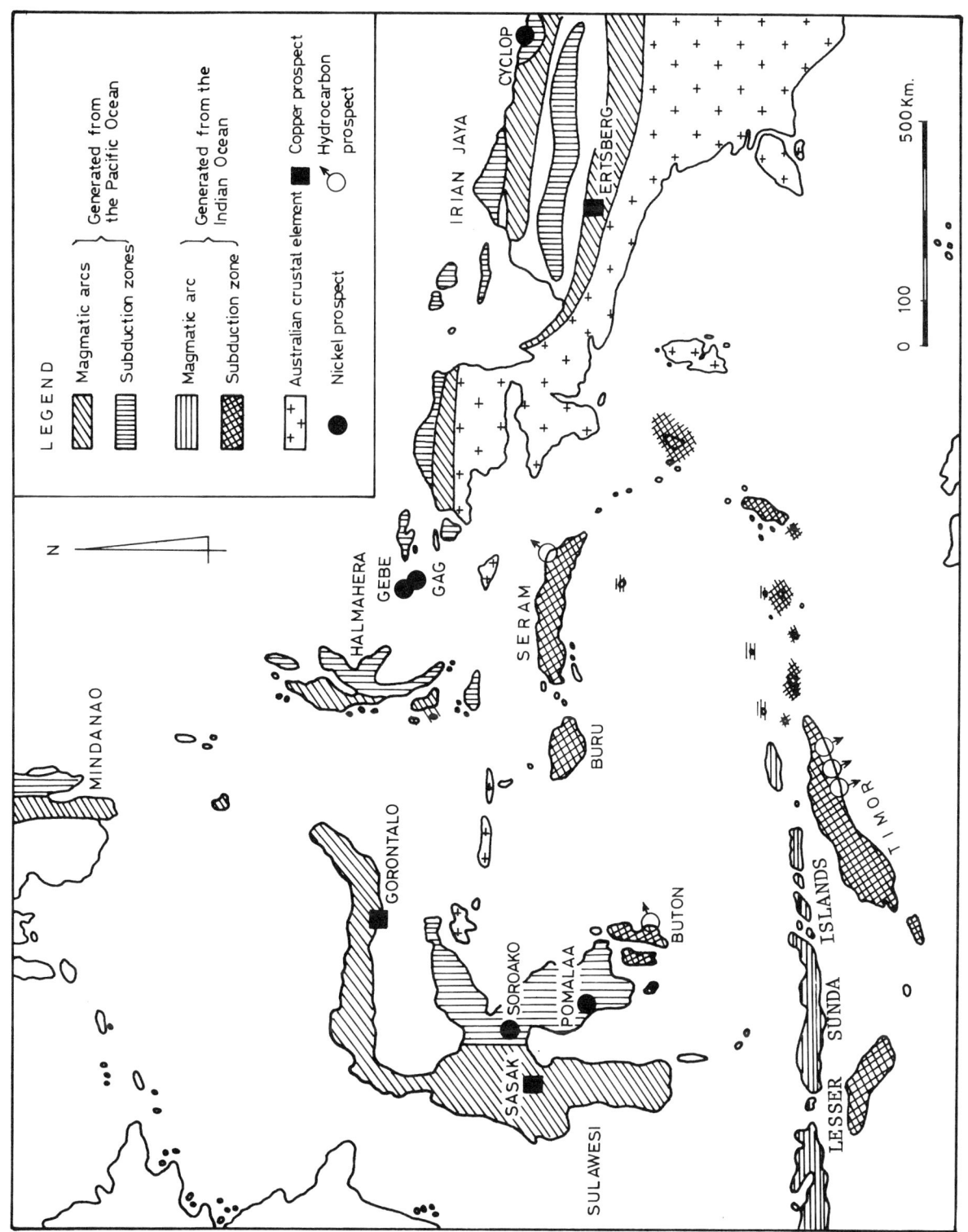

Figure 11. Subduction zones and volcanic-plutonic arcs in eastern Indonesia and their relation to mineral deposits (Katili, 1974).

deposits are located in the ensimatic arc systems of the Philippines (Taylor and Hutchison, 1978) and Sulawesi.

In general there is an excellent correlation in Indonesia between plutonic arcs of various ages envisaged by the plate tectonic model (Katili, 1974) and the occurrence of porphyry copper, manganese, gold-silver, lead-zinc-copper-silver veins, lead-zinc-copper-silver stratabound deposits, skarn deposits, and tin-tungsten-molybdenum deposits (Feisz, oral communication, 1976).

Advances have also been made in recent years regarding the genesis of tin in the light of plate tectonics. Occasionally the tin belts are paralleled by metal provinces in which other metals are dominant, such as in the orogenic belts associated with the Pacific coast of the Americas. In view of this relationship, Sillitoe (1972) proposed that the post-Palaeozoic metal provinces in these areas are related to subduction zones that were active during the Mesozoic, and the early and middle Cenozoic. Subduction is at present still locally active along the belt.

Mitchell and Garson (1976) suggested that the absence of tin from most island arcs which lack exposed pre-Mesozoic rocks provides evidence that crust of continental thickness is necessary for the generation of tin-bearing magmas. The Main Range tin granites of Malaysia are considered, as previously described by Mitchell (1977), to be the result of collision between a West Malaysia Plate and a volcanic arc lying to the east (fig. 12). However, it should be pointed out that not all granites are generated as a result of continental collision, for example those of the Himalayas which have no tin deposits associated with them.

Taylor and Hutchison (1976) argued that the major tin-tungsten and antimony mineralization in Southeast Asia is spatially associated, but not necessarily contemporaneous with acid plutonic rocks. The plutonics show clear evidence of an anatectic origin within the continental crust, and only those from the East Coast Belt of Malaysia and Billiton show any relation to the Permian-Triassic volcanic plutonic arc.

To account for the major tin deposits occurring in Southeast Asia and China, Taylor and Hutchison (1978) proposed the concept of a single Precambrian continent, rifted into separate plates by marginal early Palaeozoic basins which were welded together again by late Triassic closure of these basins. Cameron (personal communication, 1983) is of the opinion that these were separate platelets which were welded together for the first time in the Late Triassic.

Tin occurrences outside the Thai-Malaysian-Indonesian tin belt are known from the Bird Head (Irian Jaya), the island of Sula, and recently from East Kalimantan. The tin granites from eastern Indonesia could be related to fragments of the Australian continental crust that were detached from Irian Jaya by transform faulting. The occurrence of tin in East Kalimantan is associated with Tertiary plutonic rocks (?), but because the genesis of tin requires a continental setting, this occurrence could be ascribed to an old continental fragment attached to or incorporated in a young volcanic plutonic arc.

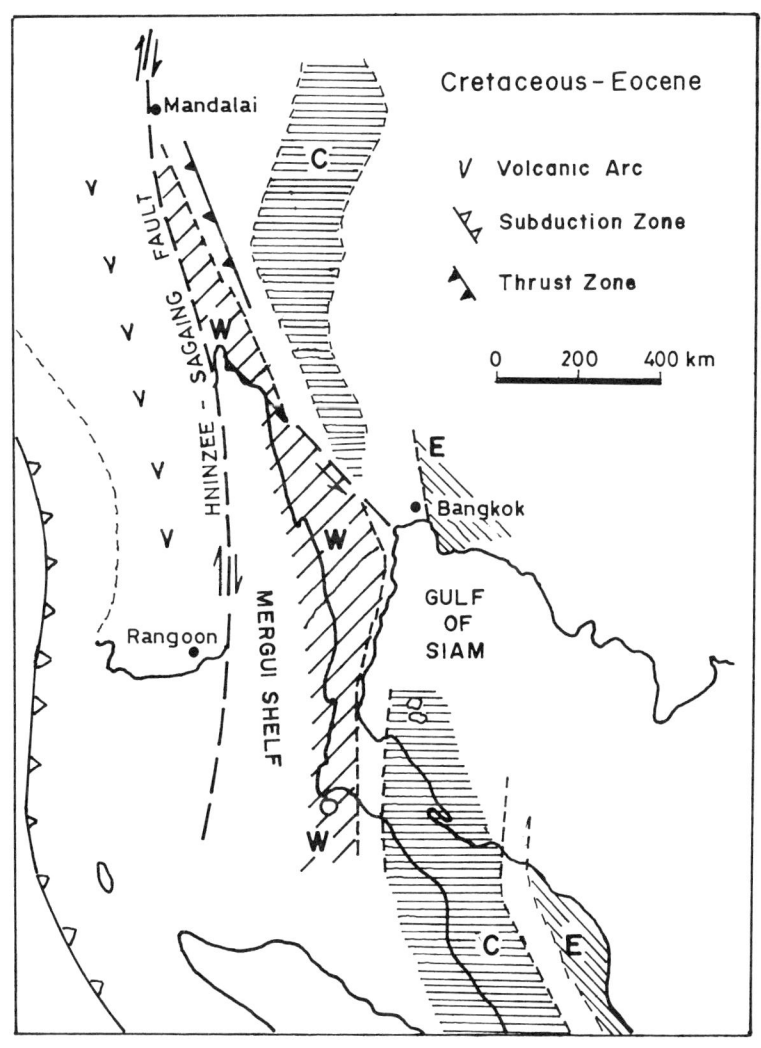

Figure 12. Southeast Asian tin province (Mitchell, 1979).

W = Western Belt (Cretaceous-Eocene back-arc magmatic
     belt tin granites)

C = Central Belt (Late Triassic collision related tin
     granites)

E = Eastern Belt (Permo-Trassic magmatic arc tin granites).

Two new classifications of calc-alkaline granites have evolved in recent years and are currently influencing exploration concepts, particularly in relation to tin in Southeast Asia. The new granitoids classification are the "I" & "S" type scheme of Chappel and White (1974) and the "magnetite" and the "ilmenite" series scheme of Ishahara (1977). According to Ishihara and others (1979), tin deposits seem to be associated with ilmenite-series granitoids, whereas the magnetic series granites lack tin. The major granitic bodies of late Palaeozoic to early Mesozoic age in southern Thailand and Peninsula Malaysia are mainly composed of the ilmenite-series type. The isolated plutons of Cretaceous to Paleogene age in this area are predominantly of the magnetite series and do not contain tin. Ishihara and others also pointed out that the typical island arc volcanic plutonic association exhibits magnetite-bearing characteristics, as demonstrated by the barren granites of the Central Intrusive Belt of the Malayan Peninsula. However, the Permian-Triassic volcanic plutonic belt of the east coast of Malaya does contain tin, and the granitoids are predominantly of the ilmenite-series type. Ilmenite-series type granites are also found in the Cretaceous volcanic plutonic arc of Sumatra, side by side with magnetite-series granites (Sujatna, personal communication, 1982). No tin occurrences have been reported so far from this region.

The relationship between porphyry copper deposits and tectonics has been treated intensively by Taylor and Van Leeuwen (1980). The Sumatra and Burma occurrences, situated along an underthrusted continental margin with young calc-alkaline volcanics, are characterized by copper mineralization and little gold, but some molybdenum. The Sumatra deposits are closely associated with the Sumatra fault system.

The copper-silver-gold porphyries of Sulawesi and the Philippines occur in areas where the underlying crust is wholly oceanic in nature and are characterized by steeply dipping Benioff zones. The Sabah porphyries occur within a major transform fault zone and seem to be unrelated to a volcanic plutonic arc or subduction zone. Cameron (personal communication, 1983) believes that copper was metasomatically removed from the ophiolites hosting the porphyries.

According to Taylor and Van Leeuwen (1980), understanding of porphyry copper deposits of Sulawesi has been enhanced by applying the model of tectonic evolution as envisaged by the present author (Katili, 1978). For most of northern Sulawesi, the geological history is that of a simple island arc built up entirely during the Tertiary from oceanic material with subduction of Pacific Plate elements along west- or northwest-dipping Benioff zones during the Miocene, Pliocene, and present day (fig. 13a and fig. 13b). Taylor and Van Leeuwen (1980) contended that porphyry mineralization occurred late in the geological history and may belong to a hiatus in the history of active volcanism related to a period of fracturing following the uplift of the earlier andesite-diorite volcanic-plutonic complex.

The same authors concluded that the Southeast Asian data reinforce the concept of close association of copper-gold porphyries with oceanic environments, and the restriction of copper-molybdenum and molybdenum deposits to more continental rock associations. The occurrence of molybdenum in Malala,

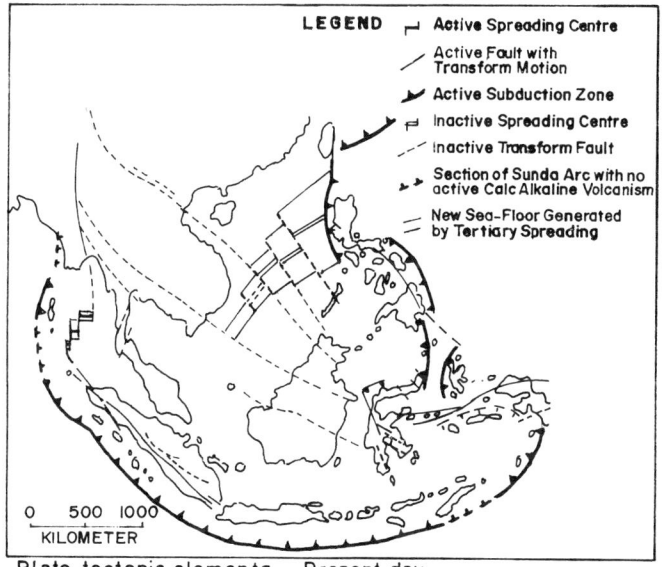

Plate tectonic elements - Present day

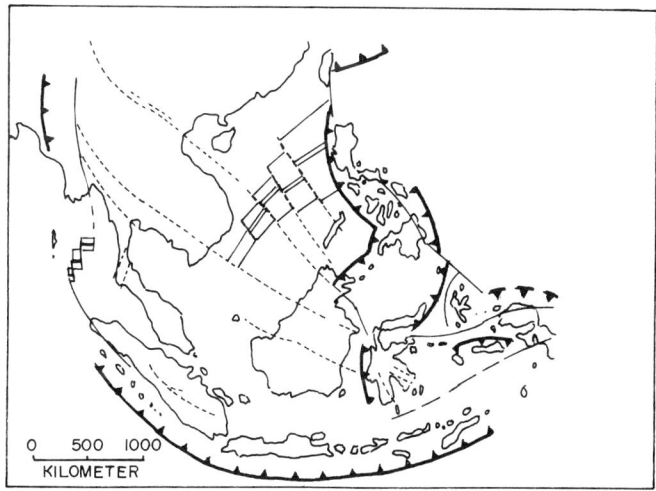

Plate tectonic elements - Upper Pliocene (1-2 m.y. B.P.)

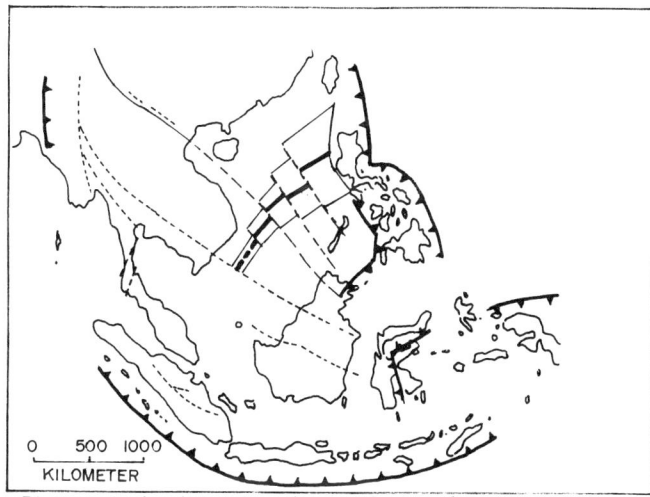

Plate tectonic elements - Upper Miocene (10 m.y. B.P.)

Figure 13a. Plate tectonic evolution of Southeast Asia from Miocene till Present (Taylor and Van Leeuwen, 1980).

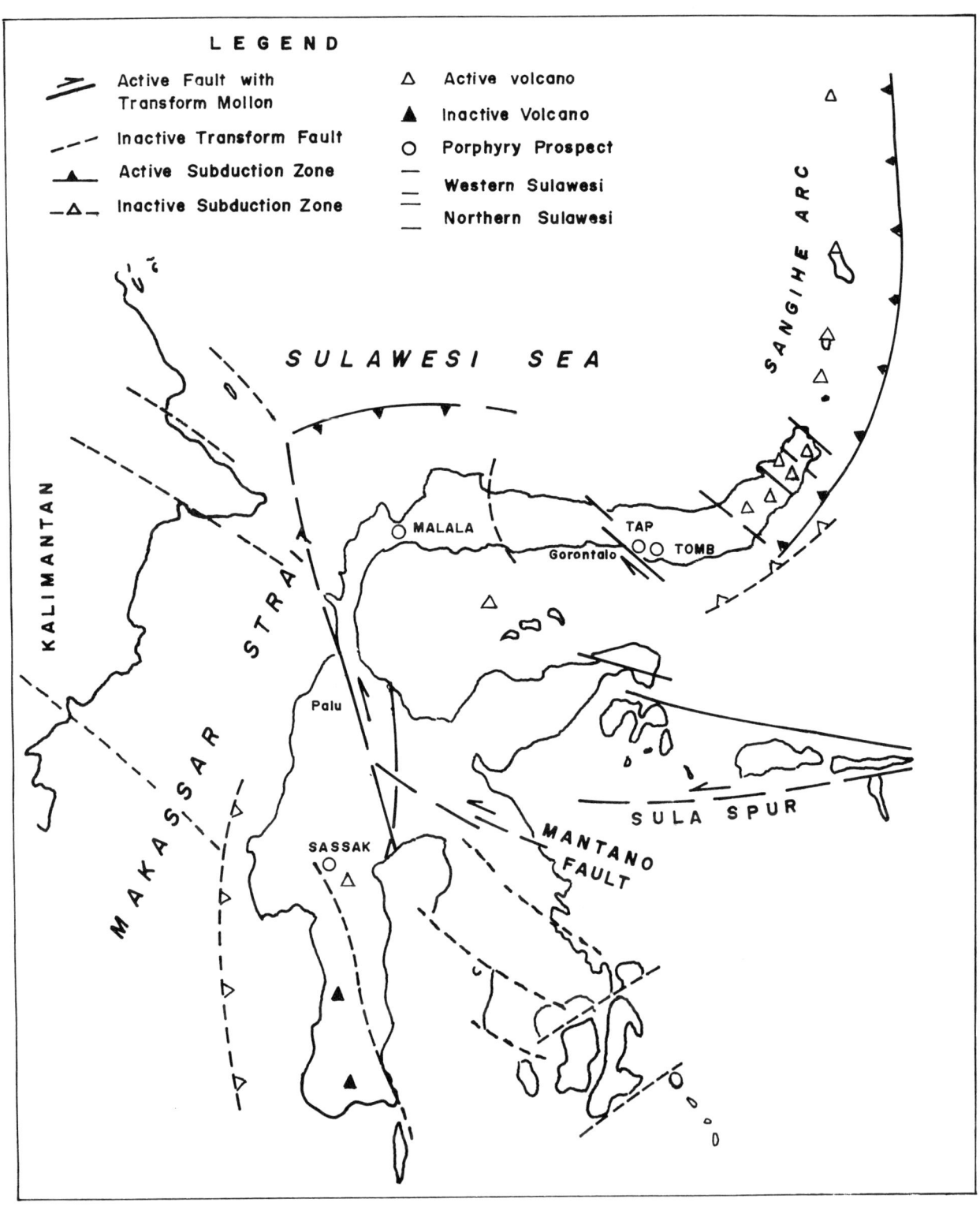

Figure 13b. Plate tectonic features of Sulawesi and porphyry deposits (Taylor and Van Leeuwen, 1980).

Central Sulawesi, in an oceanic setting suggests that these two types may be closely juxtaposed in complex tectonic environments. It may be possible that the juxtaposition of these two different tectonic settings was caused by incorporation or attachment of a small continental fragment into the Tertiary volcanic-plutonic arc.

The generally low tenor of copper mineralization in Sumatra was explained by Katili (1974) by assuming that the Indian Ocean floor was poor in copper, whereas Taylor and Hutchison (1978) suggested that the subduction was too young to have generated suitable melt. However, Taylor and Van Leeuwen (1980) are of the opinion that the fundamental reasons for the clustering of porphyry bodies in the Philippines and New Guinea and their paucity in the Sunda arc remain obscure. Uyeda and Nishiwaki (1980) suggested that the uneven and complementary occurrences of porphyry copper and massive sulphide deposits can be explained by the difference in stress fields between the Chilean-type (porphyry-copper mineralization) and the Mariana-type subduction zone (massive sulphide mineralization). As no important porphyry copper deposits and massive sulphide mineralization have been discovered so far in Sumatra and Sulawesi respectively, this concept is rather difficult to use as a basis to explain the tectonic environment of the Indonesian copper deposits.

Important mineralization in the ophiolite suite, such as nickel, occurs in Larona and Soroako, Southeast Sulawesi, the Gag-Gebe islands, Irian Jaya, and in the Nonoc island south of the Philippines (fig. 14). All these deposits seem to originate from subduction zones generated by the Pacific Ocean. As has been stated earlier, their occurrences are intimately related to collision zones and transform faults, as can be demonstrated from eastern Indonesia (fig. 15).

The important Zambales Complex of West Luzon is a slice of South China Sea crust obducted in Pliocene time (Taylor and Hutchison, 1978). The deposits consist of refractory and metallurgical-grade chromite with minor amounts of Ni-Cu sulphides. The deposit to the south of Palawan seems also to be developed in rocks derived from the South China Sea. The emplacement of these ophiolites could be the result of collision between the Reed Bank and Palawan.

Taylor and Hutchison (1978) concluded that the Southeast Asian mineral deposits may be broadly classified into two groups. The first is related both spatially and genetically to rocks that have developed in ocean basins or island arcs. The second group of mineral deposits is related spatially to continental rocks, and frequently to highly evolved granites of crustal derivation.

The evolution of these ores could be more accurately understood by taking into account such plate tectonic processes as seafloor spreading and subduction, changes in polarities of small subduction zones, time-space evolution of arc volcanism in relation to subduction (thermal process), back-arc spreading (hydrothermal activity), origin and significance of fore-arc volcanism and diapirism, arc-(micro) continent collision, shearing along transform faults, and continental accretion of allochtonous terrain, etc.

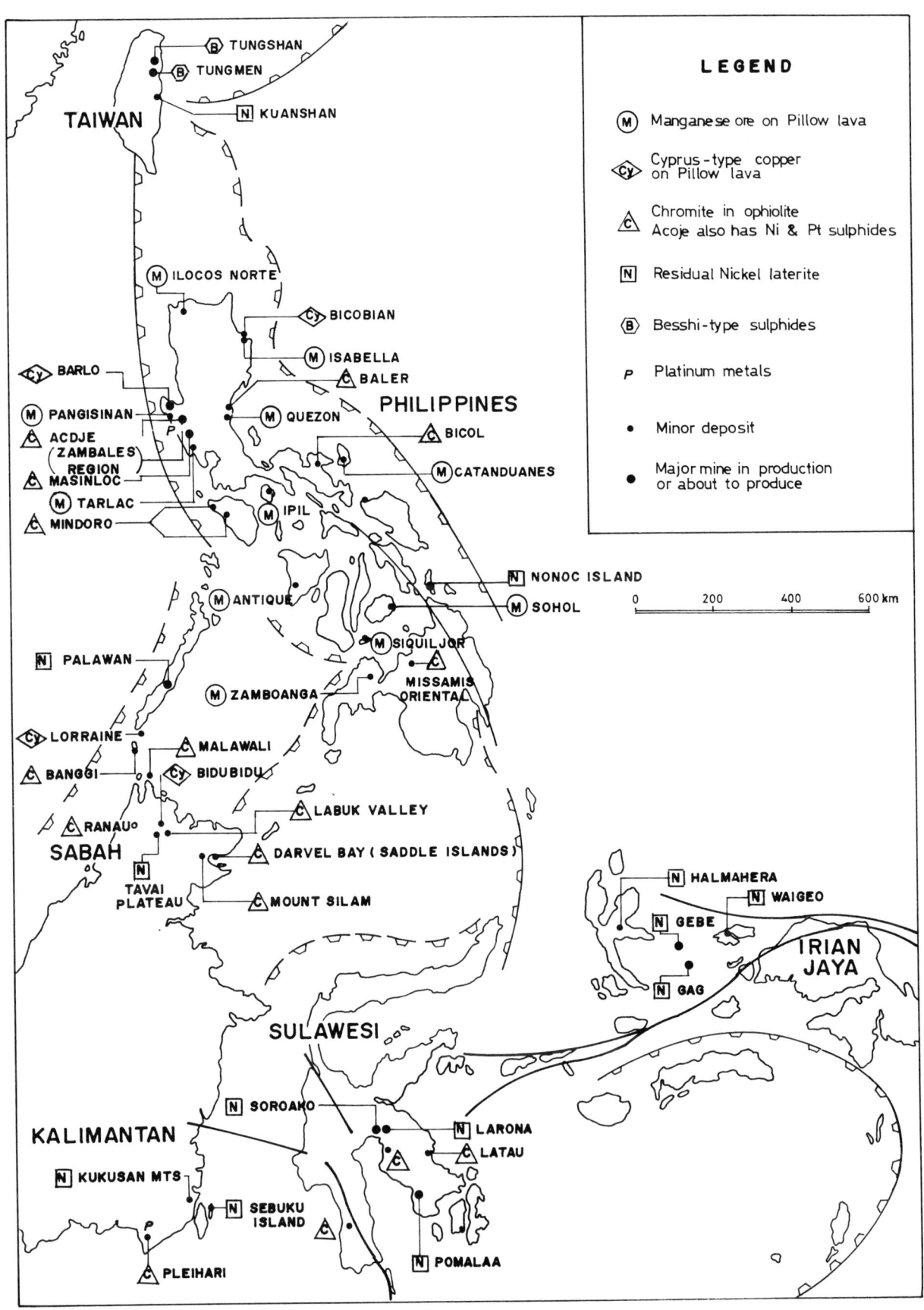

Figure 14. Locality map of ophiolite-related deposits in Southeast Asia (Taylor and Hutchison, 1978).

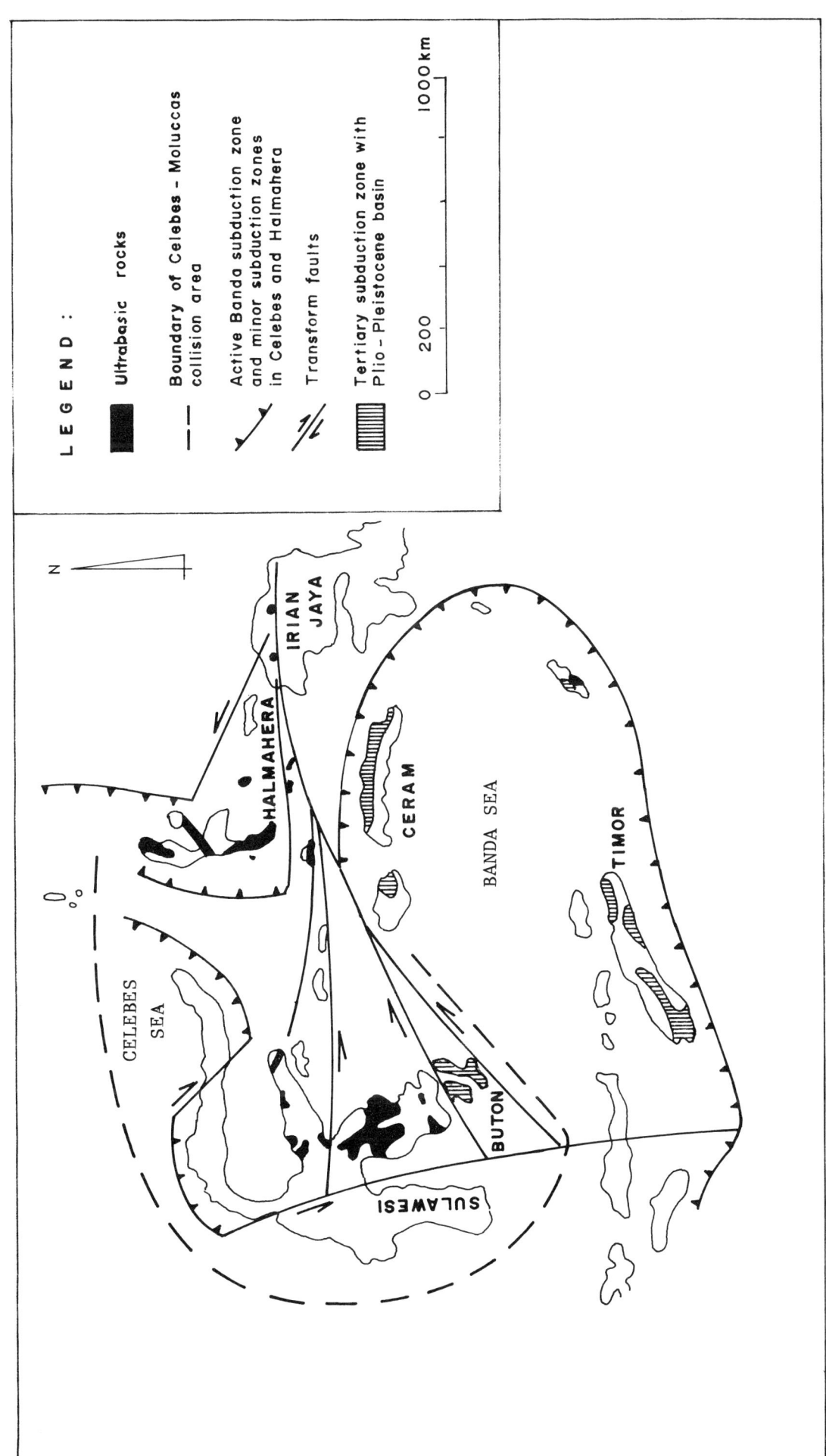

Figure 15. Eastern Indonesia showing the active and Tertiary Banda and Sulawesi subduction zones with Plio-Pleistocene basins and the Sulawesi-Moluccas collision zone characterized by ultrabasic rocks (Katili, 1975).

Basin setting and tectonic-framework

The Cenozoic tectonic framework as envisaged in this paper can also be used to gain a better understanding of the setting of the Southeast Asian oil basins. Broadly speaking Southeast Asia can be divided into three main areas: two of continental crustal affinity (Sundaland-IndoChina and New Guinea) and the oceanic island arc systems (Philippines and Sulawesi).

In discussing the tectonic evolution of Southeast Asia it has been previously argued that Sundaland has been at the core of accreting subduction systems since at least the late Palaeozoic. The subduction zones have moved systematically away from the continent towards the ocean; older subduction zones occur closer to the continent, whereas the younger ones are situated nearer to the ocean. Growth, however, has been rather asymmetric to the north and northeast.

As has been mentioned earlier, a reorganization of subduction systems occurred in Late Cretaceous-early Tertiary time. During this interval active subduction in the northeastern margin of the Sundaland ceased, namely in northwest Kalimantan and the Natuna arc. In the southeast, subduction migrated from east to southeast Kalimantan to its present position south of Java. This change in configuration of the subduction systems is considered to be very important in relation to the tectonic environment of the present oil basins.

Based on this tectonic framework, Gage and Wing (1980) proposed a classification of the oil basins of Southeast Asia which due to its simplicity, will be used throughout the discussion on the basin setting and tectonic framework of Southeast Asia. The classification as proposed by Gage and Wing is summarized in table 1:

Some of the basins will be touched upon briefly in this discussion (fig. 16).

The outer arc basin situated between the accretionary wedge and the volcanic arc such as the Mentawai basin, west of Sumatra, typically lacks the coarse quartz-rich sediments necessary for clastic reservoir formation. The heatflow is low and hydrocarbon source rocks are immature.

The Sumatran back-arc basins rest on continental crust of Mesozoic/late Palaeozoic age (fig. 17a). The area has experienced vertical mobilization and wrench/compression along its current volcanic/plutonic arc. This has resulted in the superimposition of northwest-southeast grain over the pre-existing north-south basement grain of western Sundaland. The older grain had earlier acted as control on Tertiary sedimentation. Heatflow is high and hydrocarbon source rocks are mature.

The inner-arc basin type as developed behind the Java volcanic/plutonic arc is located over the transition between continental and intermediate crust. The basin tectonic style is tensional, in the form of horst and grabens (fig. 17b). The inner-arc basin types continue northeastwards towards Southeast Kalimantan.

Table 1 : Classification of Southeast Asian Oil Basins (Gage and Wing, 1980)

SUPER-GROUPS	SUB-CATEGORIES	FAMILIES	EXAMPLES
CONVERGENCE	SUBDUCTION	OUTER ARC	Mentawai basin
		BACK ARC	Sumatra foreland basins (North, Central, South Sumatra)
		INNER ARC	Java foreland basins (NW Java, Java Sea)
		SUTURE	W. Irian, Taiwan
		INTRA OCEANIC ARC	Philippines
		INTRA/INTER BATHOLITIC	Ketungan - Melawai
		OLISTOSTROMAL/OVERTHRUST	C e r a m
	TRANSFORM	WRENCH	Tonkin Gulf
DIVERGENCE	RIFT	CONTINENTAL MARGIN (ACID CRUST)	West Natuna
	DRIFT	CONTINENTAL MARGIN (ACID CRUST)	Northwest Palawan
		OCEANIC MARGIN (BASIC CRUST)	NW Borneo (Baram), Kutai.

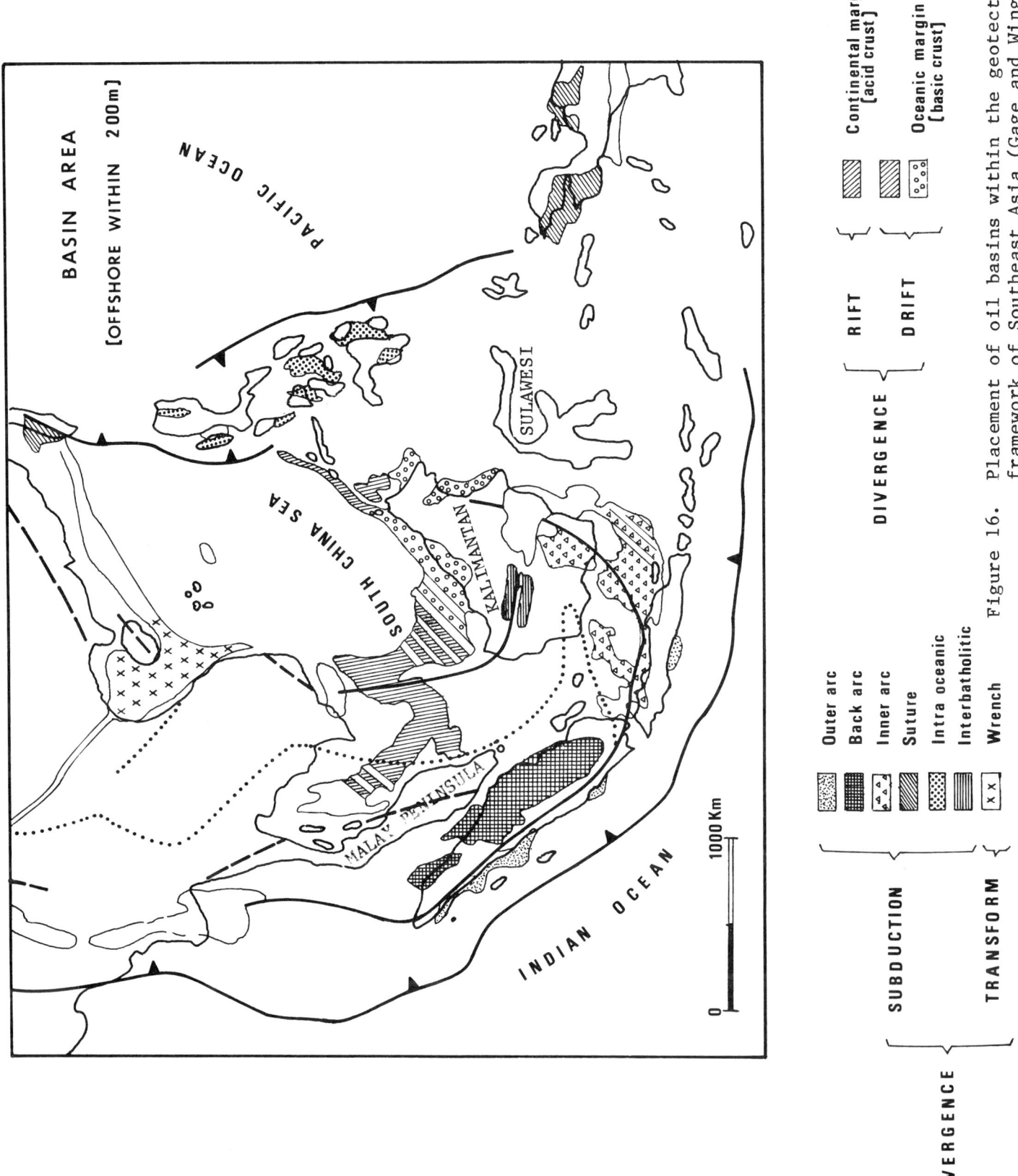

Figure 16. Placement of oil basins within the geotectonic framework of Southeast Asia (Gage and Wing, 1980).

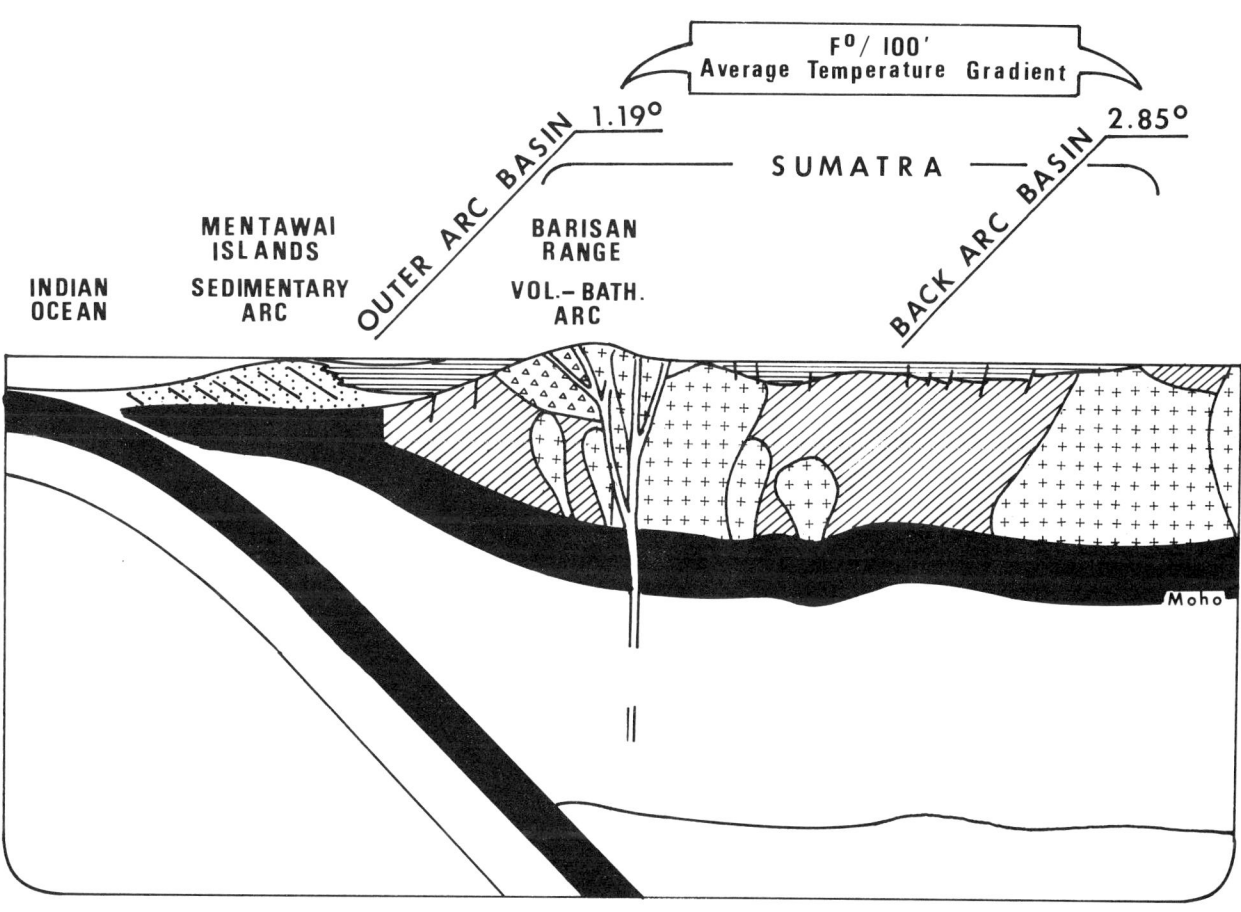

Figure 17a. Crustal section showing the Sumatran back arc basins (Gage and Wing, 1980).

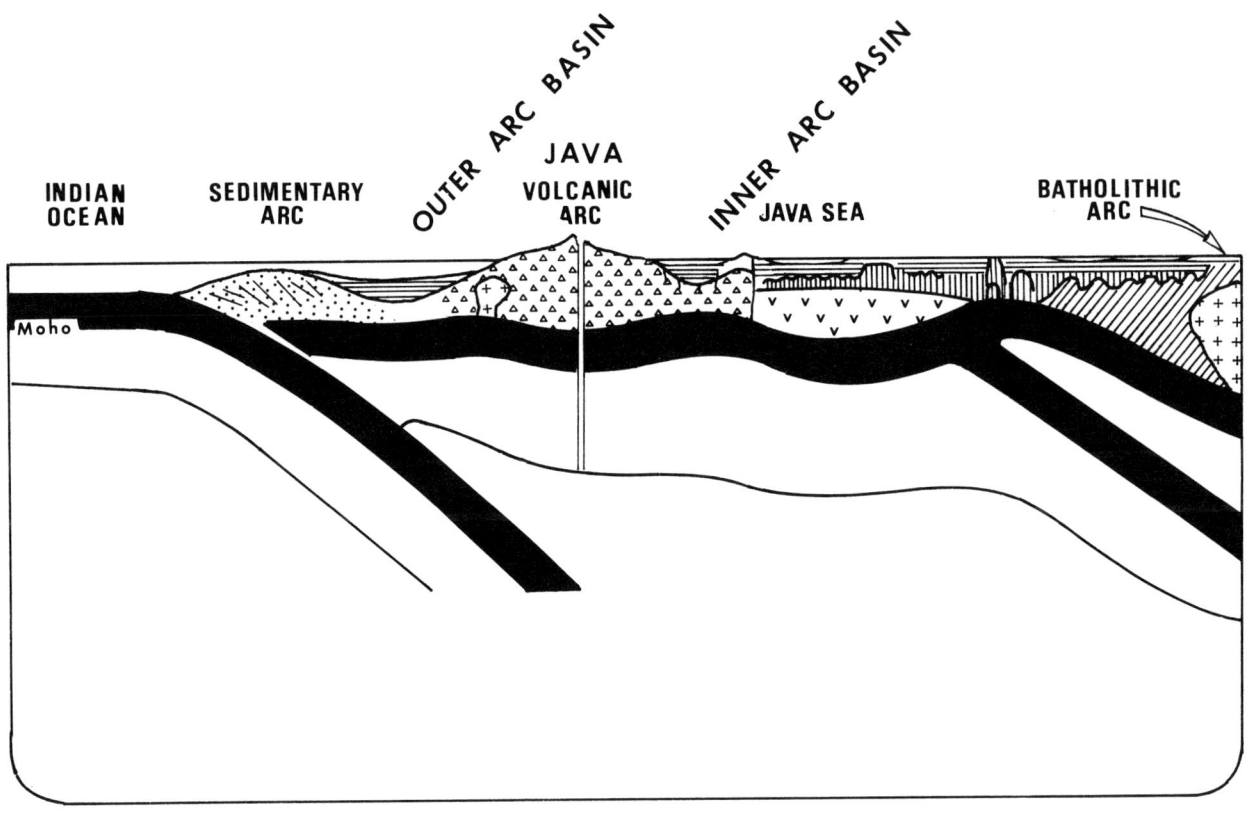

Figure 17b. Crustal section showing the Java inner arc basins (Gage and Wing, 1980).

The Sunda basin situated between Java and Sumatra is considered not to be related in the manner of outer-inner and back-arc basin types. Its axis is more or less perpendicular to these other basins and its north and north-easterly oriented grain has rifted style which most probably can be associated with the basin location at the major offset in orientation of subduction between Java and Sumatra.

The basins along the southern China shelf are of continental (drift) marginal style. Continental margin basins rank poorly in their hydrocarbon potential as a whole, but may be rich individually, such as northwest Palawan.

The West Natuna basin is considered to be a rift basin because the Oligo-Miocene spreading axis of the South China Sea continues southwards towards the axis of the West Natuna basin. Rift basins have high hydrocarbon potential owing to their often semi-restricted depositional environments and the presence of reservoirs as part of the horst and graben structures, as well as overlying draped sediments.

The Gulf of Tonkin is interpreted as a wrench-related basin lying astride the continuation of the Son-Ma-Red River fault zone. By analogy to others, these basins are favourable for hydrocarbon exploration.

The Northwest Borneo-Baram Delta and the Kutei basin have been classified as oceanic-margin basins. They are underlain by oceanic crust. Deltaic sediments are built out onto oceanic crust from continental provenances.

The tectonic setting of the various basins in the South China Sea region is depicted in fig. 18 (see also fig. 4).

GENERAL CONCLUSIONS AND RECOMMENDATIONS FOR FUTURE RESEARCH

Intensive scientific investigations in the last ten years by national and international research institutes, the outstanding contributions of the mineral industry to the body of geologic knowledge in this region and, above all, the huge amount of high quality data collected within a relatively short time, has changed Southeast Asia overnight from one of the least known into one of the well known areas in the world.

Two questions can be raised pertaining to the future of geoscientific research in Southeast Asia, namely 1) how to obtain a better balance between research studies and resource-oriented studies, and 2) how to close the gap between the extensive work done to date in oceanic areas and studies on the land portions of the SEATAR transects.

An interdependent three-stage programme of tackling these problems could be considered. This would include the following components:

a. Continuation of the basic scientific work on the tectonics of the region. This in fact was the subject of SEATAR and related studies and could be considered to have produced a very comprehensive synthesis of the broad regional tectonics. Some of the future studies might include:

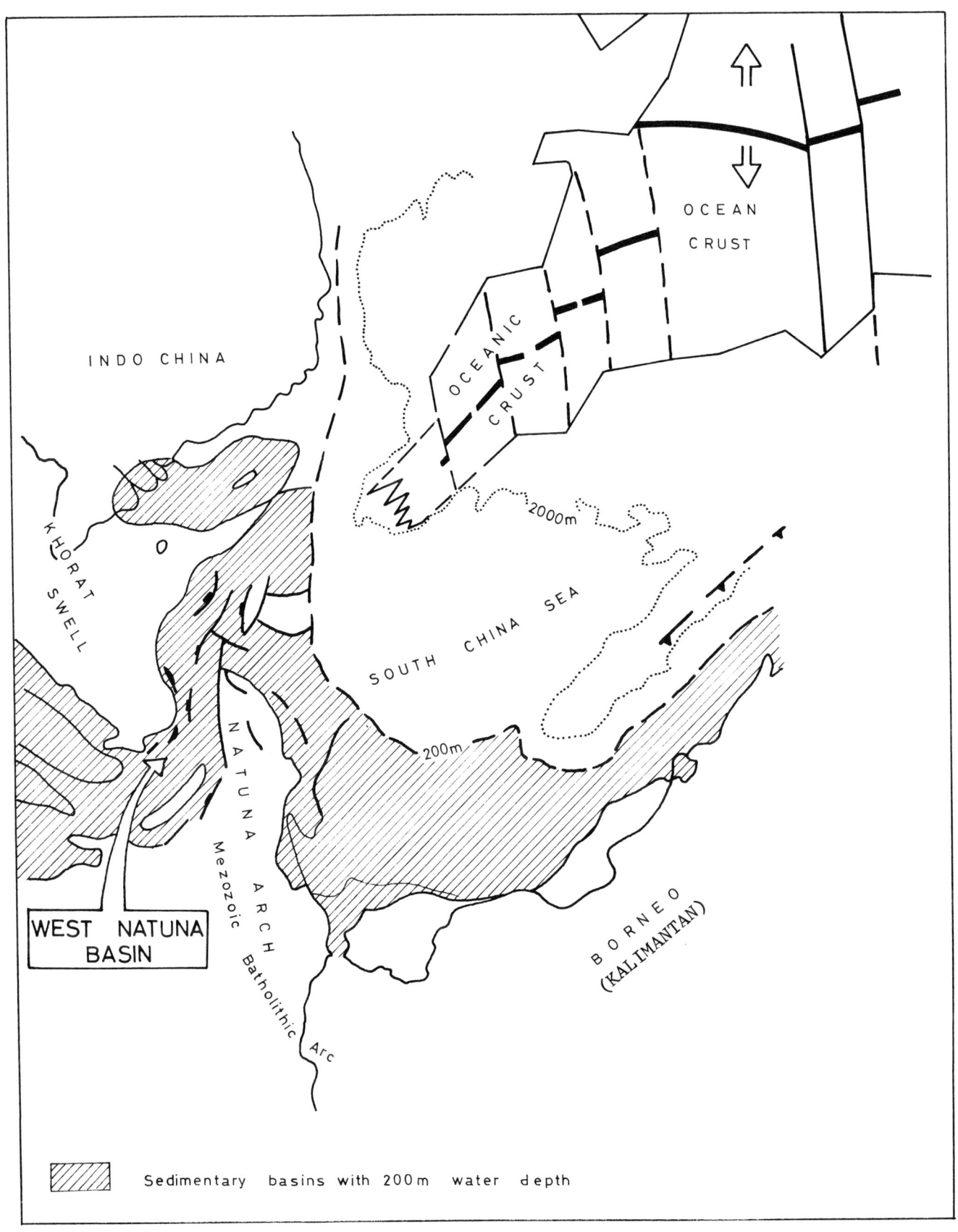

Figure 18. Sedimentary basins (200 m. water depth) of the South China Sea Region and relation of West Natuna Basin to oceanic spreader. Data from Gage (Katili, 1981).

1. The determination of the age, surface structure, and origin of the Banda Sea. The origin of the Banda Sea may provide significant insight into the kinematic processes acting during collision events. Silver and Bill (written communication, 1982) have proposed to carry out sampling preceded by multibeam swatch-mapping, high-resolution seismic reflection, and magnetic surveying of the Banda Sea. The usefullness of possible future deep drilling in this area is also being considered.

2. Study of deformation across an arc-continent collision zone. The relationships between faults, folds, volcanism, and olisostromes are critical, both for the understanding of on-going collision processes and for recognizing these processes in ancient orogenic belts. Silver (written communication, 1983) has proposed a study which will span the arc system from Timor through to the back-arc thrust zones, using as a major tool the sea MARC 2, a broad-swatch, high-resolution system capable of producing undistorted bathymetric and side-scan sonograph maps.

3. Detailed stratigraphic and tectonic studies of islands which have complicated geology, such as Sulawesi, Irian Jaya, and the Philippines, to study the role of allochthonous terranes in continental accretion and mountain building. Such studies have been initiated in the Philippines (Karig, 1982) and in Sumatra (Pulunggono and Cameron, written communication, 1983). The identification of allochthonous terranes and the description of their trajectories in time and space is a first objective, but Karig (1982) stressed that it is of fundamental importance to understand the processes responsible for the isolation, displacement, and accumulation of these elements, such as back-arc spreading and strike-slip faulting.

4. The Palaeozoic tectonic framework of the western part of Southeast Asia is not well understood. Detailed investigation of the age of the Danau formation in western Kalimantan (Borneo) and absolute age dating of granites occurring in this region are necessary. Additional geological and geophysical data from the Indosinian and China Plate margins have to be made available to the international scientific community in order to get a complete picture of the tectonic evolution of Southeast Asia, and in particular, the South China Sea basin.

5. The large transform or transcurrent faults, such as the Sumatran fault system, the Palu-Koro (Celebes), and Philippine fault, should be studied in more detail, particularly as regards their delineation, displacement, and the duration of the displacement. It is not known how the Sumatran fault system ends southeast of the island and neither do we know whether, and if so how, this movement was transformed to the Java subduction zone. The same holds for the Philippine fault zone. We do not know whether it represents an old suture or a transform fault and how it relates to the regional tectonics of the region. The Philippine fault zone, as one of the most fundamental elements in Luzon, should be examined in detail and characterized.

6. Throughout the years, scientific practice has demonstrated that the more data we acquire the more difficult it becomes to arrive at a stage where a small number of theories can account for the majority of observations. This is especially true for geochronology and paleomagnetism. The controversial results, particularly in paleomagnetism, have in certain cases created a confused picture in reconstructing the geotectonic evolution of Southeast Asia, and it is strongly suggested that these studies be carried out in a more coordinated way; correlation with the results of field geological investigations is necessary. It is not unlikely that paleomagnetic results reflect only the movement of a small allochthonous terrane within an island arc which has remained close to a "fixed" continent for several hundred million years. A closer cooperation between marine geophysicists and land-based geologists could certainly solve some of the problems postulated above.

b. A translation of the ideas and theories contained in the numerous publications on the plate tectonics of Southeast Asia into working hypothesis or schemes for relating the occurrence of hydrocarbon and mineral deposits, to the regional tectonics, which could be tested by and/or guide more detailed studies. Some examples might include:

1. The work in the Makassar Strait which, while obviously of a research nature, also provided data to assess the hydrocarbon potential. This is especially true for heat flow studies in this marginal basin. Situmorang (1982) contended that pre-lower Miocene sediments deposited at a depth greater than 2.5 km have reached a sufficient temperature for hydrocarbon generation.

2. Future research in the South China Basin pertaining to further magnetic profiling to update and enable correlation of previously linear anomalies in the central portion of the basin, seismic profiling to delineate possible areas of subsided continental fragments in the flank of the basin, and investigations of other bathymetric highs within the intermediate shelfzone for their hydrocarbon potentials in Mesozoic and Paleogene sedimentary sections.

3. The Moluccas Sea has been considered as an active arc-to-arc collision zone. Silver (1981) is of the opinion that the zone of thrusting, especially where the collision complex is carried out into the arc apron, may have significant hydrocarbon potential as it consists generally of a thick sequence of mildly folded and faulted sediments. Heat-flow information obtained in this area will certainly be of value in understanding the maturation process of hydrocarbons, as the Moluccas Sea has high biological productivity, providing an abundant source for hydrocarbon accumulation. Proximity to the volcanic arc is known to produce elevated heat flow.

4. Mineralized zones among volcanic arcs may vary according to the nature of the subducted plate, as has been suggested by Katili (1974). Position of polarity and age of subduction beneath the north and south arms of Sulawesi (fig. 13b), Halmahera, and the Philippines may be

critical in testing this hypothesis and may provide a method of prediction for the occurrence of specific ores. These studies should include identification and dating of volcanic arcs, as well as the determination of their polarity, mapping of large coherent ophiolitic bodies which may retain information regarding spreading trends, duration of spreading, etc., melange and accretionary belts in which direction of subduction and age of deformation might be preserved.

c. Applied studies based on (b) above to establish the relationship between the tectonics and resources of the region in sufficient detail to be useful in the future search for hydrocarbon and mineral resources.

In the field of hard minerals, these studies could include the proposals put forward during the Philippine SEATAR Workshop in 1981, for example the stratigraphic dating of the Kuroko deposits and the petrochemistry of the associated rocks, radiometric dating of porphyry and vein type mineralization, structural and textural investigations of chromite, and regional tectonic setting during episodes of mineralization. The results of such studies should be presented in such a way that they could be used by the governments of developing countries as a basis for policy decisions on the direction of their resources exploration.

ACKNOWLEDGEMENTS

I wish to acknowledge Mr. Hartono and his staff from the Geological Research and Development Center, Bandung, for their valuable assistance rendered to me during the preparation of this paper.

My deepest gratitude goes to Dr. Pulunggono from PERTAMINA and Dr. Cameron from CONOCO for the valuable advice they gave me in improving this article.

REFERENCES CITED

Audley-Charles, M.G., Carter, D.J., Barber, A.J., Norvick, M.S.C. and Tjokrosapoetro, S., 1981. Reinterpretation of the geology of Seram: implications for the Banda arcs and northern Australia. In The Geology and Tectonics of Eastern Indonesia, Geological Research and Development Centre, Special Publication no. 2, 1981, p. 217-237.

Bowin, C., Purdy, G.M., Johnston, C., Shor, G., Lawver, L., Hartono, H.M.S. and Jezek, P., 1980. Arc-continent collision in Banda Sea region. American Association of Petroleum Geologists Bulletin 64, p. 868-915.

Cameron, N.R., 1981. The regional tectonic setting of Sumatra. Bulletin of the Directorate of Mineral Resources Indonesia, p. 137-150.

Cardwell, R.K. and Isacks, B.L., 1978. Geometry of the subducted lithosphere beneath the Banda Sea in Eastern Indonesia from seismicity and fault plane solutions. Journal Geophysical Research 83, no. B6, p. 2825-2838.

Carter, D.J., Audley-Charles, M.G. and Barber, A.J., 1976. Stratigraphical analysis of island arc-continental margin collision in Eastern Indonesia. Journal Geological Society of London, V. 132, part 2, p. 179-198.

Chappell, B.W. and White, A.J.R., 1974. Two contrasting granite types. Pacific Geology, 8, p. 173-174.

CCOP-IOC, 1974. Metallogenesis, hydrocarbons and tectonic patterns in Eastern Asia. UNDP/CCOP, Bangkok, 158p.

CCOP-IOC, 1980. Studies in East Asian tectonics and resources (SEATAR). UNDP/CCOP, Bangkok, 256p.

Curray, J.R., Moore, D.G., Lawver, L.A., Emmel, F.J., Raitt, R.W., Henry, M. and Kieckhefer, R.M., 1978. Tectonics of the Andaman Sea and Burma. In Watkins, J.S., Montadert, L. and Dickerson, P.W., (eds.), Geological and Geophysical Investigations of Continental Margins, American Association of Petroleum Geologists Memoir 29, p. 189-198.

Eguchi, T., Uyeda, S. and Maki, T., 1979. Seismotectonics and tectonic history of the Andaman Sea. Tectonophysics, 57, p. 35-51.

Gage, M. and Wing, R.S., 1980. Southeast Asian basin types versus oil opportunities. Proceedings of the 9th Annual Convention, Indonesian Petroleum Association, p. 132-147.

Hamilton, W., 1977. Subduction in the Indonesian Region, in Island Arcs, Deepsea Trenches and Back-Arc Basins, Maurice Ewing Series, American Geophysical Union, 1, p. 15-31.

Hamilton, W., 1978. Tectonic map of the Indonesian region, U.S. Geological Survey Map I-875-D. Miscellaneous Investigations Series, 1 sheet 1:5,000,000.

Hamilton, W., 1979. Tectonics of the Indonesian region, U.S. Geological Survey Professional Paper 1078, 345p.

Hamilton, W., 1981. Plate motions in Eastern Indonesia and surrounding regions. (abstract). In The Geology and Tectonics of Eastern Indonesia, Geological Research and Development, Special Publication no. 2, p. 29.

Hayes, D.E. (ed.), 1983. The tectonic and geologic evolution of Southeast Asian seas: Part 2. American Geophysical Union Monograph 27, 402p.

Hepworth, J.V., 1982. A United Nations advisory centre for geology in Southeast Asia. Episodes, 1982/2, p. 12-15.

Hutchison, C.S., 1973. Tectonic evolution of Sundaland: a Phanerozoic synthesis. In Regional Conference on the Geology of Southeast Asia, Proceedings, Geological Society of Malaysia, Bulletin 6, p. 61-86.

Hutchison, C.S., 1975. Ophiolite in Southeast Asia. Geological Society of America Bulletin, v. 86, p. 797-806.

Hutchison, C.S., 1979. Southeast Asian tin granitoids of contrasting tectonic setting. Advances in Earth and Planetary Sciences 6, 1979, p. 221-232.

Ishihara, S., 1977. The magnetic-series and ilmenite-series of granitic rocks. Mining Geology, 27, p. 293-305.

Ishihara, S., Sawata, H., Arporusuwan, S., Busaracome, P. and Bungbrakearti, N., 1979. The magnetic-series and ilmenite series granitoids and their bearing on tin mineralization, particularly of the Malay Peninsula region. Geological Society of Malaysia, Bulletin no. 11, p. 103-110.

Karig, D.E., 1982. Accreted terranes in the northern part of the Philippine Archipelago. In Balce, G.R. and Zanovia, A.S. (eds.), Geology and Tectonics of the Luzon-Marianas Region, Philippine SEATAR Committee Special Publication no. 1, p. 67-81.

Katili, J.A., 1969. Permian volcanism and its relation to the tectonic development of Sumatra. Bulletin Volcanologique, v. 33, no. 2, p. 530-540.

Katili, J.A., 1970. Large transcurrent faults in Southeast Asia with special reference to Indonesia. Geologische Rundschau, 59(2), p. 581-600.

Katili, J.A., 1971. A review of the geotectonic theories and tectonic maps of Indonesia. Earth Science Reviews, v. 7, p. 143-163.

Katili, J.A., 1973. Geochronology of West Indonesia and its implication on plate tectonics. Tectonophysics, 19, p. 195-212.

Katili, J.A., 1975. Geological environment of the Indonesian mineral deposits, A plate tectonic approach. Publikasi Teknik Seri Geologi Ekonomi (Geol. Surv. of Indonesia), 7. Also United Nations ESCAP, CCOP Technical Bulletin, 9, p. 39-56, 1975.

Katili, J.A., 1975. Volcanism and plate tectonics in the Indonesian island arcs. Tectonophysics, 26, p. 165-188.

Katili, J.A., 1978. Past and present geotectonic position of Sulawesi, Indonesia. Tectonophysics, 45, p. 289-322.

Katili, J.A. and Valencia, M.J., 1981. Geology of Southeast Asia with particular reference to the South China Sea. Energy v. 6, no. 11, p. 1077-1091.

Katili, J.A., and Hehuwat, F., 1967. On the occurrence of large transcurrent faults in Sumatra, Indonesia. Journal of Geosciences, Osaka City University, v. 10, p. 5-17.

Kieckhefer, R.M., 1980. Geophysical studies of the oblique subduction zone in Sumatra. Ph.D. thesis, University of California, San Diego, 134p.

Klompe, Th. H.F., 1957. Pacific and Variscan orogeny in Indonesia; a structural synthesis. Madjalah Nmu Alam Indonesia (Indonesia Journal of Natural Science), v. 113, p. 43-87.

Kuenen, Ph. H., 1935. Geological interpretation of the bathymetrical results, geological results. The Snellius Expedition in the Eastern Part of the Netherlands East Indies, 1929-1930, v. 5, Part 1, 124p.

Kraus, E., 1951. Vergleichende Baugeschichte der Gebirge. Akademi - Verlag, Berlin, 588p.

McElhinny, M.W., Haile, N.S. and Crawford, A.R., 1974. Palaeomagnetic evidence shows Malay Peninsula was not part of Gondwanaland. Nature, v. 252, no. 5485, p. 641-645.

Mitchell, A.H.G., 1977. Tectonic settings for emplacement of Southeast Asian tin granites. Geological Society of Malaysia Bulletin, November 1977, no. 9, p. 123-140.

Mitchell, A.H.G. and Garson, M.S., 1976. Mineralization at plate boundaries. Minerals Science Engineering, v. 8, no. 2, p. 129-69.

Moore, G.F., Billman, H.E., Hehanusa, P.E. and Karig, D.E., 1980. Sedimentology and paleobathymetry of Neogene trench-slope deposits, Nias Island, Indonesia. Journal of Geology, v. 88, no. 2, p. 161-180.

Mrozowski, C.L. and Hayes, D.E., 1979. The evolution of the Parece Vela Basin, eastern Philippine Sea. Earth and Planetary Science Letters, v. 46, no. 1, p. 49-67.

Ninkovich, D., 1976. Late Cenozoic clockwise rotation of Sumatra. Earth and Planetary Science Letters, v. 29, no. 2, p. 269-275.

Nutalaya, P. and Rau, J.L., 1981. Bangkok: the sinking metropolis. Episodes, 1981/4, p. 3-8.

Page, B.G.N., Bennet, J.D., Cameron, N.R., Bridge, D.McC., Jeffery, D.H., Keats, W. and Thaib, J. 1979. A review of the main structural and magmatic features of northern Sumatra. In Magmatism and Tectonics in Southeast Asia, Geological Society of London Journal, v. 136, part 5, p. 569-579.

Ridd, M.F., 1971. Southeast Asia as a part of Gondwanaland. Nature, v. 234, no. 5331, p. 531-533.

Rodolfo, K.S., 1969. Bathymetry and marine geology of the Andaman Basin and tectonic implications for Southeast Asia. Geological Society of America Bulletin, v. 80, no. 7, p. 1203-1230.

Sasajima, S., Nishimura, S., Hirooka, K., Otofuji, Y., van Leeuwen, T. and Hehuwat, F., 1980. Palaeomagnetic studies combined with fission-track datings on the western arc of Sulawesi, East Indonesia. Tectonophysics, v. 64, no. 12, p. 163-172.

Sillitoe, R.H., 1972. A plate tectonic model for the origin of porphyry copper deposits. Economic Geology, v. 67, no. 2, p. 184-197.

Silver, E.A., 1981. A new tectonic map of the Molucca Sea and East Sulawesi, Indonesia, with implications for hydrocarbon potential and metallogenesis. In The Geology and Tectonics of Eastern Indonesia, Geological Research and Development Centre (Bandung), Special Publication no. 2, p. 343-347.

Silver, E.A. and Moore, J.C., 1981. The Molucca Sea Collision Zone. In The Geology and Tectonics of Eastern Indonesia, Geological Research and Development Centre (Bandung), Special Publication no. 2, p. 327-340.

Situmorang, B., 1982. The formation of the Makassar Basin as determined from subsidence curves. 11th Annual Convention, Indonesian Petroleum Association.

Smit Sibinga, G.L., 1933. The Malay double (triple) orogen. Proceedings Koninklijke Akademie Wetenschappen, Amsterdam, v. 36, no. 2-4, p. 202-210, 323-330, 447-453.

Stauffer, P.H. and Gobbett, D.J., 1972. Southeast Asia a part of Gondwanaland? (Discussion). Nature, v. 240, no. 102, p. 139-140.

Stille, H. and Latze, Fr., 1945. Die tektonische Entwicklung der pazifischen Randgebiete II. Geotektonische Forschungen, v. 7/8, Verlag von Gebrueder Borntraeger, Berlin-Zehlendorf, 323p.

Surjono and Clarke, M.C.G., 1981. Primary tungsten occurrences in Sumatra and the Indonesian tin islands. Bulletin Directoret of Mineral Resources, Indonesia, v. 1, no. 5, p. 1-40.

Tan, B.K. and Khoo, T.T., 1982. GEOSEA - its origin and development. Episodes, 1982/1, p. 9-12.

Taylor, D., van Leeuwen, T, Ishihara, S. (ed.), Takenouchi, S. (ed.). Porphyry-type deposits in Southeast Asia. In Granitic Magmatism and Related Mineralization, Mining Geology (Japan) Special Issue, no. 8, p. 95-116.

Taylor, B. and Hayes, D.E., 1980. The tectonic evolution of the South China Basin. In Hayes, D.E. (ed.), The Tectonic and Geologic Evolution of Southeast Asian Seas and Islands. American Geophysical Union Monograph no. 23, p. 89-104.

Taylor, D. and Hutchison, C.S., 1978. Patterns of mineralization in Southeast Asia, their relationship to broad-scale geological features and the relevance of plate tectonic concepts and their understanding. Proceedings of the Eleventh Commonwealth Mining and Metallurgical Congress, Hong Kong, Paper 68, p. 1-15.

Theide, J., 1983. Whither the oceanic geosciences? IUGS Commission on Marine Geology, Oslo, 163p.

Tjia, H.D., 1972. Strike-slip faults in West Malaysia. Proceedings of the 24th International Geological Congress, Section 3, p. 255-262.

Tjia, H.D., Ninkovich, D., 1977. Late Cenozoic clockwise rotation of Sumatra - comments and reply. Earth and Planetary Science Letters, v. 34, no. 3, p. 450-451.

Umbgrove, J.H.F., 1949. Structural history of the East Indies. Cambridge University Press, 64p.

Untung, M. and Barlow, B.C., 1981. The gravity field in eastern Indonesia in the geology and tectonics of Eastern Indonesia. Geological Research and Development Centre (Bandung) Special Publication no. 2, p. 53-63.

Uyeda, S. and Hales, A.L. (ed.), 1982. Subduction zones: An introduction to comparative subductology. In Geodynamics Final Symposium, Tectonophysics, v. 81, no. 3-4, p. 133-159.

Uyeda, S. and McCabe, R., 1982. A possible mechanism of episodic spreading of the Philippine Sea. Philippine SEATAR Publication no. 1, p. 53-66.

Uyeda, S. and Nishiwaki, C., 1980. Stress field, metallogenesis and mode of subduction. In "The Continental crust and its Mineral Deposits," D.W. Strangway, (ed.), Geological Association of Canada, Special Paper 20, p. 323-339.

van Bemmelen, R.W., 1949. The geology of Indonesia. Government Printing Office, The Hague, 732p.

van Bemmelen, R.W., 1978. The present formulation of the Undation Theory. Zeitschrift fur Geologische Wissenschaften, v. 6, (5), p. 523-540.

Vening Meinesz, F.A., 1934. Report of the gravity expedition in the Atlantic of 1932 and the interpretation of the results. Gravity Expeditions at Sea 1923-1932. Netherlands Geodetic Commission, Delft. v. 2, p. 47-51.

Westerveld, J., 1952. Phases of mountain building and mineral provinces in the East Indies. Proceedings of the 18th International Geological Congress. Great Britain, Part XIII, p. 245-255.

OTHER PUBLICATIONS AVAILABLE
in the
NEW IUGS SERIES

(Prices listed below are in U.S. dollars and are subject to change without notice. IUGS prices include postage and packing. Order from IUGS Offices at 601 Booth Street, Room 177, Ottawa, Ontario, Canada K1A 0E8, or Department of Geology, Norwegian Institute of Technology, 7034-NTH Trondheim, Norway.)

Publication
Number

1. Sheng Shen-fu, 1980. *THE ORDOVICIAN SYSTEM IN CHINA.* Correlation Chart and Explanatory Notes. 7p., 3 figures, 6 tables and 1 correlation chart. $6.00. ISBN 0-930423-00-3

2. Dean, W.T., 1980. *THE ORDOVICIAN SYSTEM IN THE NEAR AND MIDDLE EAST.* Correlation Chart and Explanatory Notes. 22p., 1 figure and 1 correlation chart. $6.00. ISBN 0-930423-01-1

3. Cati, F., Steininger, F.F., Borsetti, A.M., Gelati, R. (eds.), 1981. *IN SEARCH OF THE PALEOGENE/NEOGENE BOUNDARY STRATOTYPE. PART 1 - POTENTIAL BOUNDARY STRATOTYPE SECTIONS IN ITALY AND GREECE AND A COMPARISON WITH RESULTS FROM THE DEEP SEA.* Published by Giornale di Geologia, Serie 2a, Volume XLIV, Fasc. I-II, 210p., 19 plates. Available only from Giornale di Geologia, via Zamboni, 63-40128 Bologna, Italy.

5. Moreno, J.L.L. (ed.), 1981. *METALOGENESIS EN LATINOAMERICA (METALLOGENESIS IN LATIN AMERICA).* Proceedings of an international symposium February 3-6, 1980, Mexico City. In English and Spanish, 361p., 1 metallogenetic map of Mexico. $15.00.

6. Webby, B.D. et al., 1981. *THE ORDOVICIAN SYSTEM IN AUSTRALIA, NEW ZEALAND AND ANTARCTICA.* Correlation Chart and Explanatory Notes. 64p., 1 table and 1 correlation chart. $6.00. ISBN 0-930423-02-X

7. Wang Yi-gang et al., 1981. *AN OUTLINE OF THE MARINE TRIASSIC IN CHINA.* 21p., 2 tables and 1 map. $6.00. ISBN 0-930423-03-8

8. Barnes, C.R., Norford, B.S. and Skevington, David, 1981. *THE ORDOVICIAN SYSTEM IN CANADA.* Correlation Chart and Explanatory Notes. 27p., 2 tables and 1 correlation chart. $6.00. ISBN 0-930423-04-6

10. Bassett, M.G. and Dean, W.T. (eds.), 1982. *THE CAMBRIAN-ORDOVICIAN BOUNDARY: Sections, Fossil Distributions, and Correlations.* Published by National Museum of Wales, Geological Series 3, 227p. Available from National Museum of Wales, Cathays Park, Cardiff, CF1 3NP, U.K. £17 plus £1.70 postage and packing, or from IUGS for $29.00.

11. Hammann, W., Robardet, M., and Romano, M., 1982. *THE ORDOVICIAN SYSTEM IN SOUTHWESTERN EUROPE (France, Spain and Portugal).* Correlation Chart and Explanatory Notes. 47p., 1 figure and 1 correlation chart. $7.50. ISBN 0-930423-05-4

12. Ross, Reuben J. et al., 1982. *THE ORDOVICIAN SYSTEM IN THE UNITED STATES.* Correlation Chart and Explanatory Notes. 73p., 4 figures, 1 table and 3 correlation charts. $10.00. ISBN 0-930423-06-2

14. Seibold, E. and Meulenkamp, J.D. (eds.), 1984. *STRATIGRAPHY QUO VADIS?* Twelve papers from a 1982 IUGS Commission on Stratigraphy Symposium, Bad Honnef, F.R.G. Published as AAPG Studies in Geology No.16, 70p. Available from American Association of Petroleum Geologists, P.O. Box 979, Tulsa, Oklahoma 74101, U.S.A., $10.00 plus postage, or from IUGS for $11.50.

15. Wolfart, R., 1983. *THE CAMBRIAN SYSTEM IN THE NEAR AND MIDDLE EAST.* Correlation Chart and Explanatory Notes. 70p., 1 figure and 1 four-colour correlation chart. $10.00. ISBN 0-930423-07-0

16. Martinez Diaz, C., Wagner, R.H., Winkler Prins, C.F. and Granados, L.F., 1983. *THE CARBONIFEROUS OF THE WORLD: Volume 1 China, Korea, Japan and S.E. Asia.* Published by Instituto Geologico y Minero de Espana and Empresa Nacional Adaro de Investigaciones Mineras, S.A., Madrid, 245p. Available from ENADIMSA, Serrano 116, Madrid 6, Spain, or from IUGS for $20.00.